AF875887

Daniel Becker

Realtime Analysis of Large-Scale Data

Realtime Analysis of Large-Scale Data

by

Daniel Becker

KIT Scientific Publishing

Dissertation, Karlsruher Institut für Technologie (KIT)
Fakultät für Informatik

Tag der mündlichen Prüfung: 25. April 2016
Erster Gutachter: Prof. Dr. Achim Streit
Zweiter Gutachter: Prof. Dr. Henry Chapman
Dritter Gutachter: Prof. Dr. Hermann Heßling

Impressum

KIT Scientific Publishing

Karlsruher Institut für Technologie (KIT)
KIT Scientific Publishing
Straße am Forum 2
D-76131 Karlsruhe

KIT Scientific Publishing is a registered trademark of Karlsruhe Institute of Technology. Reprint using the book cover is not allowed.

www.ksp.kit.edu

This document – excluding the cover, pictures and graphs – is licensed under the Creative Commons Attribution-Share Alike 3.0 DE License (CC BY-SA 3.0 DE): http://creativecommons.org/licenses/by-sa/3.0/de/

The cover page is licensed under the Creative Commons Attribution-No Derivatives 3.0 DE License (CC BY-ND 3.0 DE): http://creativecommons.org/licenses/by-nd/3.0/de/

Print on Demand 2017 – Gedruckt auf FSC-zertifiziertem Papier

ISBN 978-3-7315-0552-5
DOI 10.5445/KSP/1000056498

Realtime Analysis of Large-Scale Data

zur Erlangung des akademischen Grades eines

Doktors der Naturwissenschaften

von der KIT-Fakultät für Informatik
des Karlsruher Instituts für Technologie (KIT)

**genehmigte
Dissertation**

von
M. Sc. Daniel Becker
aus Ehringshausen

Tag der mündlichen Prüfung:	25. April 2016
Erster Gutachter:	Prof. Dr. Achim Streit
Zweiter Gutachter:	Prof. Dr. Henry Chapman
Dritter Gutachter:	Prof. Dr. Hermann Heßling

Abstract

Upcoming facilities in photon science offer entirely new research opportunities for scientists. For example, it will be possible to create three-dimensional images of objects at the nano scale. Although the amount of data taken by the detector devices will increase drastically, only a fraction of the data can be used in subsequent analyses. Inherent limitations in the experimental setup result in a huge amount of empty or meaningless images being taken. The aim of this thesis is to develop algorithms for selecting suitable data and thereby make such experiments feasible in the first place.

Nanocrystallography explores the structure of macromolecular objects such as proteins. X-rays are used to create diffraction images from crystallized samples. Once diffraction images from many different orientations have been captured, it is possible to reconstruct the spatial structure of an object. The samples are transported by a liquid stream which crosses a laser beam generated by an X-ray source. A detector is used to take images of the diffracted light. The X-ray source generates flashes of light at a fixed rate. Due to technical limitations, it is not possible to synchronize the stream of samples with the laser flashes. This results in many useless images containing no information at all or the amount of signals within an image is too small for further research.

At the Linac Coherent Light Source (LCLS) experiment, 120 images per second are captured, all of which are stored offline and are analyzed later on concerning their suitability for further analysis in photon science. However, this will not be an option in the next generation of experiments. The 'European XFEL' experiment for example will be able to take up to 27,000 images per second.

In this thesis, strategies for handling this large amount of images are explored. We introduce a neural network, which is able to separate useless from useful images successfully, provided the amount of noise in an image is not too high. In addition, we propose an algorithm able to detect diffraction information within images. We are able to identify most of the signals within an image compared to the software currently in use. This is done by noise removal and edge detection followed by signal localization. We also indicate why current state of the art noise removal algorithms cannot be used in nanocrystallography.

The algorithms presented in this thesis are designed to be executable in parallel. We discuss the impact of parallel execution on reducing the data stream in photon science as close to the detector as possible.

To explore our algorithms they are implemented as a prototype. Different quantities such as efficiency and runtime behavior are studied.

Zusammenfassung

Kommende Einrichtungen für Photon Science eröffnen Forschern neue Möglichkeiten. Es wird möglich sein, dreidimensionale Bilder im Nano-Bereich zu erstellen. Obwohl die Menge der erfassten Daten rapide steigen wird, kann durch prinzipielle Einschränkungen nur ein Bruchteil dieser für Analysen verwendet werden. Durch diese Einschränkungen werden eine große Menge nutzloser Bilder aufgenommen. Ziel dieser Arbeit ist es, Algorithmen zu entwickeln, um die verwertbaren Daten auszuwählen und somit die Datenflut einzudämmen.

Nano-Kristallographie erforscht den Aufbau makromolekularer Objekte wie Proteine. Röntgenstrahlen werden verwendet, um Beugungsmuster von kristallisierten Proben zu erzeugen. Mit Hilfe von vielen Aufnahmen mit verschiedenen Ausrichtungen des Kristalls ist es möglich, die räumliche Struktur des Objekts zu rekonstruieren. Die Proben werden mit Hilfe einer Flüssigkeit transportiert. Diese Flüssigkeit führt die Proben durch einen Röntgenlaser. Dieser erzeugt Lichtblitze mit einer festen Frequenz. Aus technischen Gründen ist es nicht möglich, diese Lichtblitze mit dem Strom der Proben zu synchronisieren. Das führt dazu, dass viele Bilder entweder gar keine Informationen, oder zu wenige für eine weitergehende Untersuchung enthalten. Am 'Linac Coherent Light Source' (LCLS) Experiment werden 120 Bilder pro Sekunde erzeugt. Alle Bilder werden gespeichert und später auf ihre Eignung für weitere Analysen untersucht. Dieses Vorgehen wird in der nächsten Generation von Experimenten nicht mehr möglich sein. Das 'European XFEL' Experiment ist beispielsweise in der Lage, 27.000 Bilder pro Sekunde aufzunehmen.

Im Rahmen dieser Arbeit werden Strategien zum Umgang mit diesen Datenmengen untersucht. Wir stellen ein neuronales Netz vor, das in der Lage ist, verwertbare von nicht verwertbaren Bildern zu unterscheiden, sofern das Grundrauschen nicht zu groß ist. Außerdem stellen wir einen Algorithmus vor, der in der Lage ist, Beugungsmuster innerhalb eines Bildes zu erkennen. Verglichen mit der zur Zeit verwendeten Software, sind wir sind in der Lage, einen Großteil der Signale innerhalb eines Bildes zu erkennen. Hierzu wird zunächst das Rauschen verringert und eine Kantenerkennung durchgeführt. Im Anschluss werden Signale lokalisiert. Wir zeigen außerdem, warum aktuelle Algorithmen zur Rauschentfernung nicht in der Nano-Kristallographie verwendet werden können.

Die Algorithmen, die in dieser Arbeit vorgestellt werden, wurden entwickelt um parallel ausgeführt zu werden. Wir diskutieren den Einfluss der Parallelität auf die Reduktion des Datenstroms in der Photon Science nahe an der Datenquelle. Um unsere Algorithmen zu untersuchen sind diese prototypisch implementiert. Aspekte wie die Effizienz und das Laufzeitverhalten werden hier analysiert.

Acknowledgments

I would like to thank Prof. Dr. Achim Streit for supervising and reviewing my thesis, for guidance throughout my studies, and for always having a friendly ear for questions and concerns.

I also thank Prof. Dr. Hermann Heßling for co-supervising and reviewing my thesis, the constant stream of constructive feedback, and solution-oriented approach.

I would also like to thank Prof. Dr. Henry Chapman for reviewing my thesis.

Thanks to my colleagues Parinaz Ameri, Uğur Çayoğlu, Max Fischer, Arsen Hayrapetyan, Dr. Christopher Jung, Eileen Kühn, Marco Strutz, and Frank Tristram, for providing feedback on my thesis as well as all the fruitful discussions throughout our meetings. Furthermore I want to thank Dr. Anton Barty as well as Dr. Steve Aplin for providing me with background information and data on the physics part of my research.

In the same way I am thanking the LSDMA project for financing my research and making this work possible in the first place. In this context I also want to thank the leader of the DSIT group, Dr. Marcus Hardt.

In addition I would like to thank my parents and friends for supporting and encouraging me throughout my studies. In particular I want to thank Christopher Baynes for providing feedback on my work as well as the fruitful discussions and new perspectives on problems, Anna Döge for providing a helping hand with illustrations, Peter Große for providing new perspectives on issues, Andreas Parisek for providing feedback on my work, Tom Renker for the fruitful discussions, and Jan Thöle for providing his perspective on my work.

Contents

List of Figures

List of Tables

Chapter 1

Introduction

The amount of data taken in industry as well as science is constantly increasing. Even though the computational landscape is evolving, the handling of large scale data produced by businesses or scientific experiments is still a challenge. In recent years the term big data [63] has been established to describe this phenomenon.

Kaisler et. al [54] describe big data as a constantly evolving object:

> As little as 5 years ago, we were only thinking of tens to hundreds of gigabytes of storage for our personal computers. Today, we are thinking in tens to hundreds of terabytes. Thus, big data is a moving target. Put in another way, it is that amount of data that is just beyond our immediate grasp, e.g., we have to work hard to store it, access it, manage it, and process it.

As a result of this definition, big data describes data that can not be processed using traditional computer systems and software. In theory, it is possible to process such data in any desired way, given a sufficient amount of time. However, this is not feasible under real world circumstances, since scientific research or business decisions depend on timely conclusions drawn from the analysis of these data. Therefore, processing of the data needs to be done efficiently to provide applicable results in a reasonable amount of time.

This problem is currently surfacing in the scientific community of photon science. For a long time it was possible to carry research data around on a flash drive and process it on a personal computer. But with new generations of experiments and detectors able to take more data in shorter periods of time, this is not feasible anymore. The next logical step is to move storage to data centers and process data on local servers. However, eventually the software developed to analyze the data does not scale well enough. This leads to new challenges, residing in the big data domain.

One approach to achieve the required scalability is parallel processing. Big data frameworks such as Hadoop [93], or Spark [95] are used to distribute the data over multiple servers. Each portion of the data may then be processed in parallel using map reduce [39] or other applicable algorithms. However, this method only speeds up the processing under specific conditions, since the data analyzed has to be fully available and distributed beforehand [55]. In addition, the computational resources to store and analyze the data scale at least linearly with it. This limits the ability to quickly draw conclusions on new data. Finally, having a large fraction of meaningless data increases processing times and storage requirements.

Therefore, it is desirable to reduce the amount of unnecessary data as much as possible in order to conserve storage space, network/disk bandwidth, and computational resources.

This thesis has been made in affiliation within the 'Large-Scale Data Management and Analysis' (LSDMA) project [89]. LSDMA bridges the gap between data acquisition and analysis by combining community specific support with common cross community development. The goal of LSDMA is to streamline community specific tools for a more general scientific audience. The project is organized in sub-projects. Five of them are called 'Data Life Cycle Labs' (DLCL) and aim to support specific communities such as Climatology or Energy. The sixth sub-project is called 'Data Services Integration Team' (DSIT). An overview of the structure of the project can be found in Figure 1.1.

DSIT provides "[...]generic technologies and services for multi-community use based on research and development in the areas of data management, data access and security, storage technologies and data preservation." [89]. It is further partitioned into six work packages. This thesis contributes to the DSIT work package 'Data-intensive computing'. In specific, the work in this thesis is aimed at X-ray nanocrystallography [65].

In X-ray nanocrystallography the atomic structure of macromolecules is analyzed. To gain insights into their structure, samples are crystallized and then illuminated by X-rays. The diffracted light is then captured by a detector device. This image represents a slice through the sample in Fourier space. See Chapter 2 for more details.

By combining many images with different orientations, a three-dimensional model of the Fourier coefficients can be determined. With the coefficients, the spatial structure of the macromolecule can be reconstructed [77]. It is important that the raw data provided by the detector cannot be used directly for determining the spacial structure of objects. Instead, data need to be transformed from Fourier space to real space. This particular aspect is important for noise reduction.

Due to limitations in the experimental setup, up to $95\,\%$ of the images captured by the detector are either blank or inadequate for further analysis [25]. Currently, all images taken are stored for later offline analysis. This is a waste of storage and computational resources. For current experiments, such an approach is feasible due to the low data rate of $525\,\mathrm{MBps}$.

However, this approach is no longer feasible for new generations of experiments. The European XFEL for example is designed to achieve up to 27,000 images per second [56]. As of today, there are no solutions available to efficiently deal with this volume of nanocrystallography images. However, taming the flood of data already during the initial phases of the data taking seems to be feasible in nanocrystallography by taking its characteristics properly into account. Solutions have to be developed reducing the amount of data online and as close to the detector as possible.

This leads to the main research question of this thesis: **Is it possible to design an algorithm capable of rejecting all images which are useless for further research within the real-time-constraints of current as well as next generation experiments?**

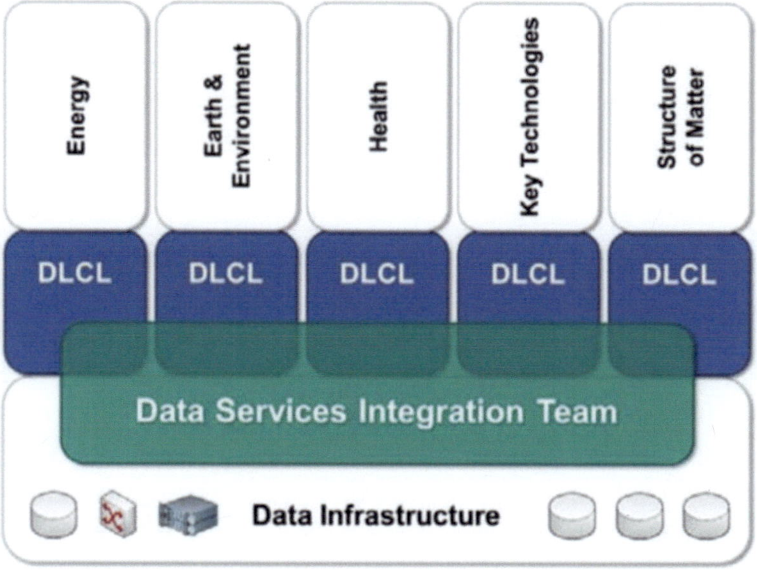

Figure 1.1: Structure of the Large Scale Data Management and Analysis Project (LSDMA) [89]. Community specific data lifecycle labs (DLCLs) doing joint research and development. They are supported by the data services integration team (DSIT), providing generic methods research and development, creating interfaces between federated data infrastructures and individual communities [9].

1.1 Research Questions

The main research question can be further split into the following sub-questions:

1. **How is it possible to determine if there is adequate data within an image at all?**
   The most basic decision an algorithm has to make is the distinction between images containing useful data and images not containing useful data. The output of the computation should be a binary information whether an image is useful or not.
2. **Is the data within an image useful for further analysis?**
   Given an image does contain data, the next question is, whether the data in the image is sufficient for more complex analysis.
   Experience has shown that an image needs to contain at least 20 valid signals to be useful to current algorithms in X-ray nanocrystallography [17].

3. **Is it possible to solve the previous two questions within real-time constraints?** Given enough time, it is easily possible to thoroughly analyze every image for data. Nevertheless, this is not feasible given real-time constraints. This means that algorithms suitable for online analysis have to meet time constraints given by experiments.

4. **Can existing algorithms be used or adapted to facilitate the image optimization?** In traditional image processing, there are plenty of algorithms for removing noise in images. The methods whereupon these algorithms rely have to be reevaluated critically with special attention to the particularities in nanocrystallography.

1.2 Publications

Within the scope of this thesis, the following articles have been published:

1. Daniel Becker and Achim Streit: "A Neural Network Based Pre-Selection of Big Data in Photon Science". IEEE Fourth International Conference on Big Data and Cloud Computing (BdCloud), 2014. IEEE, Pages: 71 - 76, DOI: 10.1109/BDCloud.2014.42, 2014, 3.-5.12.2014, University of Technology, Sydney

2. Daniel Becker and Achim Streit: "Localization of Signal Peaks in Photon Science Imaging." UKSim-AMSS 17th International Conference on Modeling and Simulation, 2015. (Best Paper Award), IEEE, Pages: 296 - 301
DOI: 10.1109/UKSim.2015.35, 25.-27.3.2015, University of Cambridge (Emmanuel College)

3. Daniel Becker and Achim Streit: "Real-time Signal Identification in Big Data Streams Bragg – Spot Localization in Photon Science". The International Conference on High Performance Computing & Simulation (HPCS 2015), Pages 611 - 616 DOI: 10.1109/HPCSim.2015.7237101, 20.-24.7.2015, Amsterdam

4. Daniel Becker and Achim Streit: "Real-time Signal identification in Photon Science Imaging". International Journal of Simulation Systems, Science & Technology, IJSSST Volume 16-3, DOI 10.5013/IJSSST.a.16.03.01

5. D. Becker, A. Streit, "Realtime-Processing of Nanocrystallography Images", Proceedings of the 18th UKSIM-AMSS International Conference on Modeling and Simulation,(Best Paper Award), IEEE, 2016, pp. 190-195, DOI: 10.1109/UKSim.2016.20, 5..-8.4.2016, University of Cambridge (Emmanuel College)

1.3 Scientific Contributions

In this section, previously listed publications are discussed in detail. Each contribution is detailed and associated with the research question it answers. We first introduce a neural network able to categorize images from nanocrystallography efficiently. Then, an algorithm is shown able to identify individual signals within images. We also show why it is not feasible to use most state of the art algorithms for noise removal in photon science. Finally, a prototypical implementation of the proposed algorithms is introduced.

1.3.1 A Neural Network Based Pre-Selection of Big Data in Photon Science

Neural networks are explored in terms of their applicability as a veto engine in photon science. A small neural network is introduced able to categorize up to 93 % of the data correctly, given the signal-to-noise ratio is sufficiently high. To improve the signal-to-noise ratio, two optimizations are introduced. Background subtraction calculates an average photon background which is subtracted from images before the analysis. In addition we identify the 'transverse intensity' as a new quantity in nanocrystallography. It defines a compensation factor for the loss of intensity at the outer areas of the detector. By applying the transverse intensity, signals at the outer area of the detector are increased.

This contribution addresses research question 1, discussed in the previous section. It will be presented in detail in Chapter 5.

1.3.2 Localization of Signal Peaks in Photon Science Imaging

Individual signals within images from nanocrystallography are identified. An algorithm composed of three steps is introduced. Firstly, single pixel noise is removed using convolution. Secondly, edge detection is applied to enhance the signals. Thirdly, signals above a given threshold are located. Up to 90 % of signals within an image are identified compared to software currently in use.

Research question 2 is addressed by this contribution. It is presented in detail in Chapter 6.

1.3.3 Real-time Signal Identification in Big Data Streams Bragg – Spot Localization in Photon Science

In our previous work we identified noise as the main challenge for the correct categorization of images as well as identification of signals. This article analyzes the 'Block-Matching 3D' (BM3D) algorithm. It is a state of the art algorithm for noise removal. In the article, its applicability for images from nanocrystallography is discussed. It is shown, that these types of noise removal algorithms do not work for this specific kind of images. This is due to the very similar shape of noise and signal.

This contribution addresses research question 3 and will be presented in detail in Chapter 6.

1.3.4 Realtime-Processing of Nanocrystallography Images

In this article, the algorithms introduced in Section 1.3.1 and 1.3.2 are implemented as a prototype. First, the implementation is discussed. Then, the prototype is explored in terms of its efficiency and runtime behaviour. The efficiency is compared agains software currently used for identifying signals in images and categorizing them. Finally, based on the benchmarked runtime, we extrapolate the results to propose a solution for handling the real-time demands of the European XFEL experiment.

Research question 3 is addressed by this contribution. It will be presented in detail in Chapter 7.

1.4 Structure of the Thesis

The remaining part of the thesis is structured in the following way:

Chapter 2 (Data Capturing in Nanocrystallography) introduces the context of research. First, nanocrystallography is introduced. Then, a typical experimental setup is described, and, finally the data flow of this experiment is laid out.

Chapter 3 (State of the Art) discusses the current state of data analysis in nanocrystallography. Technologies and algorithms used in this theses are introduced here.

Chapter 4 (Problem Description) discusses the challenges introduced by a new generation of experiments generating much more data. Furthermore, research question 4 is dealt with here.

Chapter 5 (Neural Network as a veto engine) deals with research question 1. A data categorization engine is introduced which is capable of classifying data from nanocrystallography into 'useful' and 'not-useful'.

Chapter 6 (Signal Identification) introduces a multi-step algorithm capable of identifying signals within an image. This chapter deals with research question 2.

Chapter 7 (Performance Aspects) proposes a prototypical implementation of the previously developed algorithms. The implementation is then discussed in terms of its implementation details, recognition rate as well as runtime behavior. Here, research question 3 is dealt with.

Finally, Chapter 8 draws a conclusion from the previous chapters and tries to give an outlook on further research opportunities.

Chapter 2
Data Capturing in Nanocrystallography

In this chapter the scientific area of nanocrystallography is introduced. First, Bragg diffraction is discussed, which represents the foundation of nanocrystallography. Then, nanocrystallography itself is introduced. Finally, a typical experimental setup as well as the process of data acquisition and analysis is presented.

(a) Diffraction of waves by a single slit [2]. Green represents no change in amplitude, red a positive amplitude and blue a negative one. Plane waves passing the slit are transformed to circle waves. The change in the wave pattern are a Fourier transform of the obstacle encountered by the waves.

(b) Diffraction of waves by two slits [2]. Green represents no amplitude, red a positive amplitude and blue a negative one. The waves form two cylindrical patterns when passing the two slits. When those patterns overlap with each other, the amplitudes (negative and positive) are either increase or canceled out.

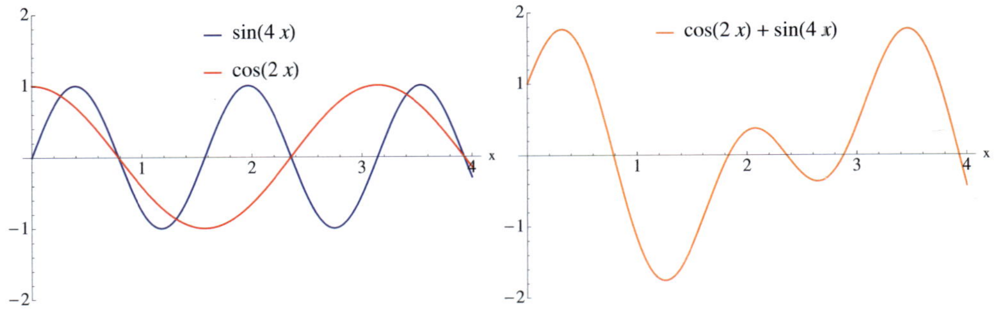

(a) $\sin(4x)$ (blue) and $\cos(2x)$ (red). **(b)** Superposition of the functions $\sin(4x)\cos(2x)$.

Figure 2.2: Plot of two signals and the sum of both signals (b). When the signals are added, amplitudes are either increased or cancel each other.

2.1 Bragg diffraction

In order to understand the research area of nanocrystallography, diffraction must be understood first. The general principle of diffraction can be well illustrated by slit experiments.

Bragg diffraction describes the diffraction of light by a three-dimensional periodic structure such as molecules in a crystal, and was discovered by Bragg et al. [26]. Bragg diffraction uses X-rays, i.e. light of very short wavelength. Due to the superposition property of light waves, sharp intense signals can be detected. They are known as Bragg spots. Based on the distribution and intensities of the Bragg spots, the spatial structure of the molecules can then be reconstructed, essentially by a Fourier transform [14], the technical details are beyond the scope of this thesis.

Due to the superposition property of waves, interference phenomena can be observed when light propagates around an obstacle or through an opening. In diffraction physics, the wavelength of the light is small compared to the dimensions of the obstacles and openings. The right hand side of Figure 2.1a illustrates how a plane wave passing through a slit is transformed into a circle wave. Similar effects can be observed when a pointlike obstacle is hit.

The superposition principle of waves is best illustrated in one dimension. Figure 2.2a shows the plots of two waves of different wavelengths. The second part of the Figure shows the result of adding the amplitudes of both waves. Depending on the amplitudes of both input signals, the resulting signal at $x = 0.5$ as well as $x = 1.25$ is doubled whereas at $x = 2$ the signals are almost canceled.

Figure 2.1b shows plane waves hitting an obstacle containing two slits. As the two resulting circle waves meet, they interfere with each other. Depending on the distance of both slits to a certain point, the phases either cancel each other, increase or, decrease. This creates a distinct diffraction pattern. From the resulting diffraction pattern the widths of the slits and their distance can be reconstructed. The reconstruction is in principle, what is done in nanocrystallography [84].

2.2 Nanocrystallography

X-ray light sources have been used for a long period of time in order to determine the inner structure of molecules [94]. The light sources have been evolving at a very high pace since their discovery. Within only 121 years, the peak brilliance [14] has been increased from 10^7 to 10^{34} (European XFEL [56]). X-ray light sources have been utilized for many significant discoveries in biology [4]. The discovery of the bulk of protein structures has been facilitated by these light sources. However, the remaining structures are more challenging. This is due to different attributes like their size and structure [30]. The common workflow here is to illuminate a crystallized sample using an X-ray light source. The light diffracted by certain planes in the crystal is then captured by a detector as an image. The signals are also called 'Bragg spots' [26]. Images are taken with different orientations of the sample. These images are then combined into a three-dimensional model of average Fourier coefficients for each coordinate x, y, z. An example of this model can be seen in Figure 2.4. Here, spheres of different diameter represent the intensity of a discrete coordinate in the Fourier space.

Based on this model, the electron density of a sample can be reconstructed. Figure 2.5 shows the spacial structure of the photosystem I protein complex obtained from LCLS diffraction data [31].

The intensity of X-rays required to create a diffraction pattern destroys a sample within femtoseconds [51]. In order to be able to determine the inner structure of these macromolecules, femtosecond X-ray nanocrystallography has been developed [31]. Femtosecond X-ray nanocrystallography uses multiple samples of the same macromolecular type to capture diffraction images from various orientations. The samples are transported through the interaction point with the X-ray light source via an automated probe delivery system, which uses a transportation liquid or gas to move the samples [90]. The light source works with a fixed repetition rate at which it generates very bright light flashes. The resulting diffraction information of the sample is then captured by a detector right before the probe is destroyed. A typical setup of such an experiment is shown in Figure 2.3.

Since a sample is floating freely in the transportation liquid, its orientation is not known. Therefore, before the data can be combined into a three-dimensional model, a process called indexing has to take place. The indexing process determines the orientation of the crystal by calculating the lattice coordinates and comparing it to a geometrical construct (Ewald's sphere, [44]). Using this derived orientation, or index, the images are then grouped for use in next steps of the analysis chain. The main challenge is the identification of valid Bragg spots.

Firstly, due to the automated setup of the experiment, it is not possible to synchronize the stream of probes with the X-ray flashes. Therefore, it is likely that a probe will not be hit at its center, resulting in weak Bragg spots. It may even happen, that no sample is hit at all, which will result in an empty image. Secondly, noise is added to the images by diffracted light from the transportation liquid.

The current generation of experiments, e.g. the 'Linac Coherent Light Source' (LCLS), are able to generate 120 images per second [83]. However, the efficiency, which we define as

$$\text{efficiency} = \frac{\text{indexable images}}{\text{images taken}} \tag{2.1}$$

can be as low as $4.5\,\%$ due to the challenges just mentioned [25]. To reduce the overhead of indexing all images, a pre-selection is carried out to limit the indexing process to images containing sufficiently strong Bragg spots. The pre-selection at the LCLS experiment for example is carried out using the Cheetah software (see Section 3.1). The data is categorized in probably indexable (useful) and probably non-indexable (useless) images. The images are captured by a detector device sensible to X-rays. It is discussed in detail in the following section.

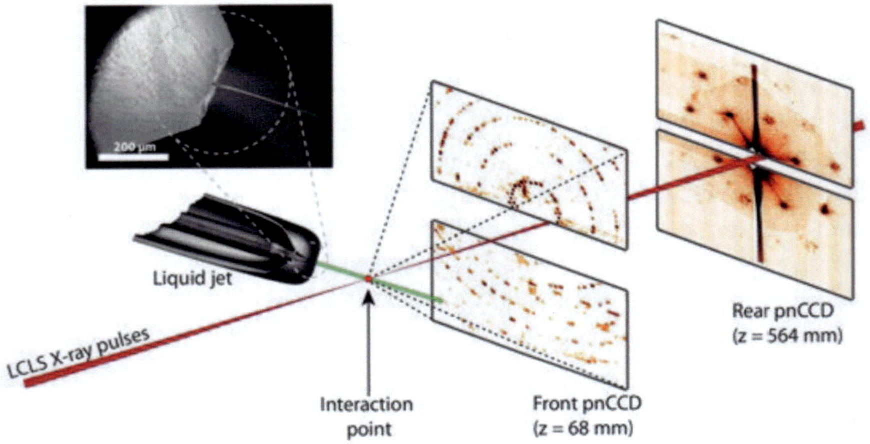

Figure 2.3: Experimental setup for serial femtosecond crystallography [31]. An X-ray light source produces bright flashes at a fixed rate. Probes are transported through the interaction point by a liquid. The diffracted light is captured by two detectors in different distances.

Figure 2.4: Three-dimensional view of the collected Fourier-coefficients [11]. For each coordinate x, y, z the combined discrete diffraction intensities of many individual images are shown. Spheres of different diameter represent the intensity of a discrete coordinate in the Fourier space

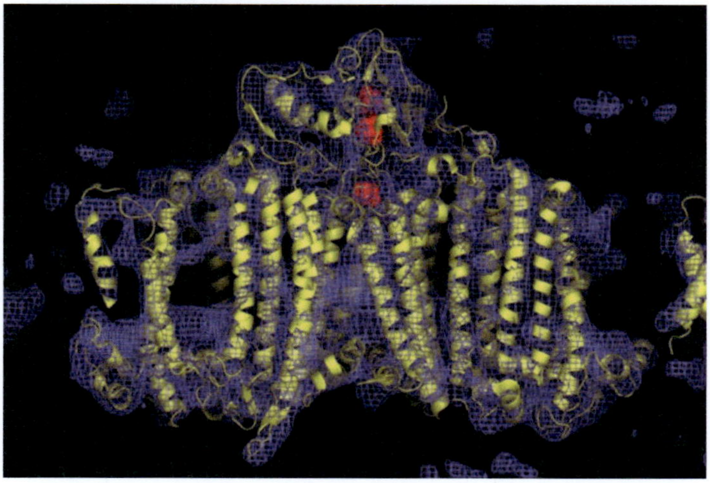

Figure 2.5: The electron density map of the photosystem I protein complex obtained from LCLS diffraction data [31].

2.2.1 Detector

The diffraction information generated within the experiments are captured by specialized detectors. At the LCLS experiment, the 'Cornell-SLAC Pixel Array Detector' (CSPad) is used [47]. For each experimental installation, the detector has to be calibrated and tested, before it is able to consistently capture data [29] [28]. An image of the detector surface can be seen in Figure 2.6.

The detector is composed of four quadrant plates, allowing the beam aperture to change in order to align the beam as well as optimize the angle to capture as many diffraction information as possible. On each quadrant eight basic elements called '2x1' are mounted. Each 2x1 module is composed of two 'Application-specific integrated circuits' (ASICs) connected to the sensory surface. For each quadrant, the corresponding ASICs are connected to an analog board providing adjustments to the sensitivity of the pixels. In addition, they contain a 'Field-programmable gate array' (FPGA), which takes care of controlling the ASICs and sends the data to an aggregation system using a 625 Mbit link. The data is aggregated by a custom data acquisition system, merging information of all four quadrants and adding further information about the recorded image. It also provides an online display of the data recorded. In this step, it is possible to plug in modules to filter the data or carry out more sophisticated analysis.

The images taken by the detector are 14 bit greyscale. Its pixel values are stored as a 1552×1480 matrix in a proprietary format.

2.2.2 Detector Geometry

Since the detector is composed of many individual sensors, the data taken by the individual panels are stored in a 8×8 matrix with one row and column of zero values around each panel. As a result, the real world coordinates of individual pixels of the detector are not static. They may vary for different experiments and even for individual runs within the same experiment. To calculate the correct position for each pixel, a geometry file is produced for each run, describing the offset relative to the captured position.

Pixels within each panel are square and have a size of 110 µm. Pixels at boundaries between two ASICs on the same panel are larger, rectangular and have a width of 275 µm. See Figure 2.8. The stored geometry of the individual panels can be seen in Figure 2.7 in the raw format as well as with the geometry file applied.

Figure 2.6: CSPad detector used at the LCLS experiment [3]. It consists of 64 individual panels mounted on for quadrants. The quadrants can be slid towards and away from the center in order to capture different diffraction angles.

(a) Panel order stored in file.

(b) Geometry applied

Figure 2.7: An image captured by the detector is not stored in its physically correct arrangement. Panels are stored as an 8×8 matrix in the form shown in (a). To rearrange the panels, the corresponding geometric information has to be applied as done in (b).

2.3 Dataflow

Currently, all data taken by the detector are stored offline for later analysis. They are stored in a proprietary data format specific to the experiment. An illustrated view of the flow of data can be seen in Figure 2.9. In a first analysis step, all images are processed by the 'Cheetah' software (see Section 3.1), which identifies and locates Bragg spots within each image. All images for which Cheetah is able to identify a certain amount of Bragg spots are then exported to the hdf5 file format along with a protocol of the Bragg spots found, their intensities, and coordinates. Those pre-selected, most likely indexable images are then used for the actual analysis process and the reconstruction of the electron density of the sample using for example the CrystFEL software (see Section 3.2).

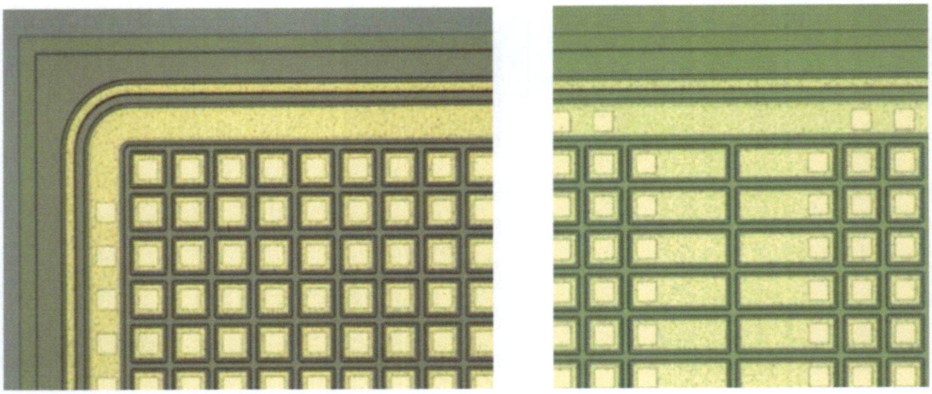

Figure 2.8: Left pixels within a panel are shown. They are square and have a size of 110 µm. On the right pixels at the boundaries between two panels on a 2x1 module are shown. They are larger, rectangular and 275 µm wide [47].

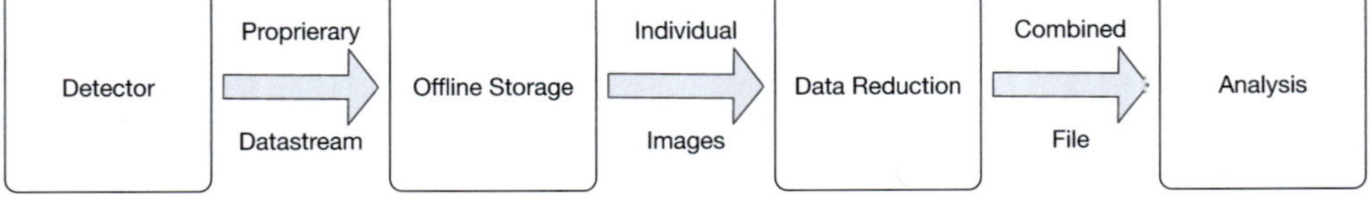

Figure 2.9: Flow of the data taken at the LCLS experiment. First, all data are stored offline. Then a data reduction is taking place, only exporting useful images for further analysis.

Chapter 3
State of the Art

In this Chapter the current state of the art is presented. In the beginning the 'Cheetah' and 'CrystFEL' software currently in use for processing nanocrystallography images are introduced. Then, articles related to our work are discussed. The Chapter is concluded by the introduction of algorithms and technologies used in this thesis.

3.1 Cheetah

The Cheetah software has been developed for analysis and data reduction of images from femtosecond X-ray nanocrystallography [17]. It was written at the Center for Free-Electron Laser Science (CFEL) by experts in that field of research. The software was developed as a universal library in C++. Since there is no default data format for experiments taking diffraction images, Cheetah can be imported into existing software, which is capable of reading the experiments specific data format. Cheetah in turn outputs individual images as tables in the Hierarchical Data Format 5 (HDF5, see [45]) container format. The main functionality of Cheetah is the search for Bragg spots within an image. In order to achieve the best recognition rate, a series of optimizations is performed.

- Broken or stuck pixels can be marked as faulty and will be ignored during the analysis.
- Based on a separate file containing geometric information about the detector, panels are analyzed separately.
- On the basis of a series of empty images, an average is calculated for each pixel of the detector. This average will be subtracted in the following analysis process reducing the background noise.
- For each element of the detector, a gain factor is calculated, which is subtracted from each pixel of the image before the actual analysis.

After applying the optimizations, Cheetah starts with the actual analysis process. The user has to supply an intensity threshold as the minimum intensity for data to be considered a signal. During the analysis, Cheetah iterates over each pixel, looking for those with an intensity above the specified threshold. In case such a pixel is found, adjacent pixels are analyzed until no more adjacent pixels with intensity above the set threshold are found. If the amount of pixels in the spot found is within pre-defined margins and if the signal-to-noise ratio of the number of those pixels and those in their surrounding is sufficiently high, the signal found is recorded as a Bragg spot. In addition, a center as well as a total intensity is calculated for that spot. Moreover, all pixels the spot is composed of are marked as processed and will be ignored from that point on.

Based on the experience of the developers, an image is marked as useful for further analysis if it contains at least 20 Bragg spots [17].

Cheetah has been successfully used to reduce the amount of data in the determination process of many different structures. For example for determining the 'Structure of a photosynthetic reaction centre determined by serial femtosecond crystallography' [53] and 'Serial femtosecond crystallography of G protein–Coupled receptors' [62].

3.2 CrystFEL

CrystFEL is a toolkit developed at CFEL [92]. Its purpose is to determine the orientation of a crystal from its diffraction pattern. It is also able to combine multiple diffraction patterns into a single three-dimensional model. In addition, diffraction patterns can be simulated for testing purposes. CrystFEL uses HDF5 as the format for input data. Therefore, it is independent of proprietary data formats generated by various detectors. Internally, CrystFEL uses a generic model of a detector. Actual detector modules can be mapped to its abstract internal detector by supplying a file containing the geometric information of the detector used to take the data.

During the process of the analysis different components play a role. First the module 'indexmajiq' is used to determine the orientation of a crystal based on the diffraction pattern captured. This process is called indexing. Similar orientations are grouped. In order to determine the orientation, multiple algorithms can be chained and will be tried until one yields a reasonable orientation. If no algorithm is able to find an orientation, the image is discarded as not-indexable. Once indexmajiq has processed all images, 'process_hkl' then combines all intensities recorded into one three-dimensional model containing the intensities of all discrete coordinates x, y, z. This component is also able to scale possible variances of the intensities to a normalized one to compensate for irregularities of detector pixels.

CrystFEL is in active development. Features and fixes are constantly added. The developers further promise to be able to handle the diffraction patterns of multiple crystals within one image, which is currently not possible. If they succeed, the concentration of crystals in the transport stream could be increased and thus yield more useful data in a shorter period of time [92].

3.3 Application of a neural network in high-throughput protein crystallography

In the article 'Application of a neural network in high-throughput protein crystallography' [23], A. Berntson et al. explore the applicability of neural networks to ensure the quality of crystal growth based on their diffraction patterns. The article states that in order to produce good diffraction images containing intense spots, the crystal has to be of high quality. Therefore, the researchers infer that based on the spot intensity and the slow decrease of intensity towards the outer areas of the detector, the crystal quality can be assessed. As a criterion for a signal, its mean intensity has to be at least twice the standard deviation of the image. The minimum acceptance size for a spot is 5 pixels. Using this criterion, images are divided into 14 concentric rings relative to the center of the detector. For each ring, spots are detected. Based on the number of spots, the distribution within the rings and the intensity distribution, a neural network is trained. Once the neural network is trained, it outputs a numerical value indicating the crystal quality. Therefore, an expert has to be involved, defining a threshold above which a result is considered useful. The findings were implemented in the open-source software 'CrySys', which was developed in C++.

3.4 Automated diffraction image analysis and spot searching for high-throughput crystal screening

The article 'Automated diffraction image analysis and spot searching for high-throughput crystal screening' [96] by Zhang et al. presents a software package called 'DISTL', able to rapidly analyze X-ray diffraction patterns. The goal is to provide information to optimize growth and cryoprotection of the crystals, freeing researchers from the need to manually inspect images except for those, the software reports as deserving special attention. To achieve this, strong Bragg spots are located. Here, a valid spot consists of a number of connected pixels with an intensity above a given threshold and is located outside of area with high noise. The location of the spots found are validated against a model in order to ensure they are valid. Finally, the quality of each spot is gauged by its size and shape. In addition, the entire image is evaluated in terms of the overall noise, quality and distribution of the Bragg spots. After all spots are found, the quality of each is rated. The quality of a spot is quantified by its size, peak intensity, shape, number of maxima and presence of nearby spots. The software package is implemented as a library, which can be included in existing analysis software. For each image passed to the library, a list of signals found is returned along with the calculated quantities for each spot. Due to the thorough analysis of the image, the authors state a typical response time for a 10 MB image of 2.5 s on a 2.8 GHz Intel CPU.

3.5 Crystalline object evaluation by image processing

In the article 'Crystalline object evaluation by image processing' [24] the usage of techniques from image processing is explored to verify the growth of crystals in X-ray microscopy. To produce diffraction images of high quality, the crystals grown for the analysis have to be of high quality. Attributes like temperature, chemicals to be added as well as the type of protein affect the crystallization process. Even though automatic crystallization systems exist, experts have to verify the crystals grown by them. Only the presence or absence of crystals can be measured automatically. The authors propose an algorithm to automatically rate the growth of the crystal utilizing the camera built into the automatic crystallization system. Firstly, the pictures taken by the system are converted to greyscale. Secondly, the Sobel operator [85] is applied to the image to detect edges. The image is then converted to a binary image, where all pixels with an intensity above a given threshold are set to 1, and the rest is set to 0. Thirdly, the actual features used for classification are extracted. The selected features are the longest line segment and the number of line segments in an image. For the decision process, a support vector machine (SVM) is used as a classifier. The SVM is able to categorize $86.3\,\%$ of the images correctly compared to an expert. The challenge remaining is the dynamic determination of a sensible threshold for the binary image conversion. Currently, a reasonable value is found through trial and error. This value is specific to the current crystallization system.

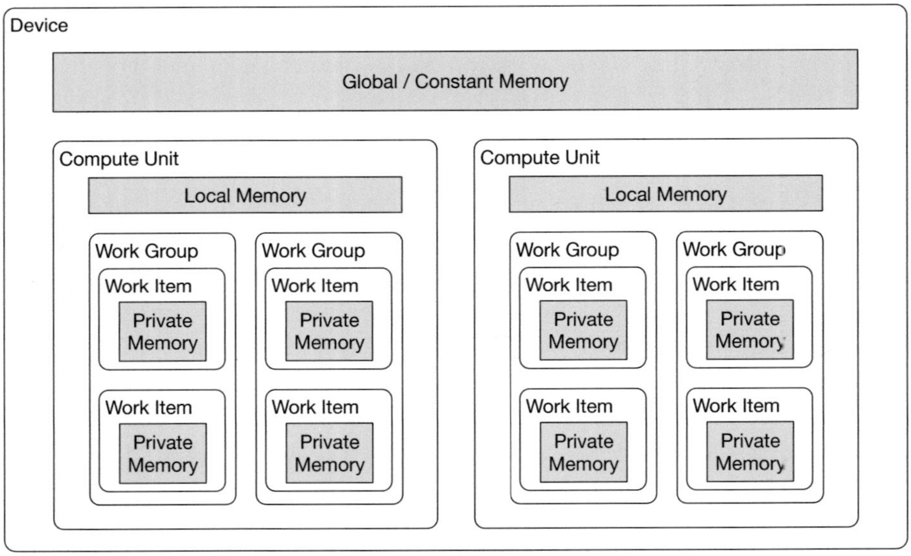

Figure 3.1: OpenCL device model. Each device has global memory. It is composed of one or more compute units. Each compute unit has its own local memory. Tasks within each compute unit are grouped into work groups and have their own private memory [87]. Genera lly speaking, the closer the memory is located to the work item, the faster it can be accessed.

3.6 OpenCL

The Open Computing Language (OpenCL) is a framework designed to unify the development of software that is capable of running on a multitude of hardware accelerators such as GPUs and FPGAs as well as CPUs. These are uniquely treated as platforms [80]. OpenCL defines its own programming language based on C99 as well as an API to control the execution of code on the actual platforms. In order to achieve platform independency, OpenCL code has to be loaded and compiled at runtime into so called ‘program kernels’. OpenCL provides task-based as well as data-based parallelism. Different tasks can run in parallel on heterogeneous hardware or the same task can run across multiple devices using different datasets. OpenCL defines a memory hierarchy for each computing device.

It is composed of

- global memory – shared across the entire device, has high latency.
- read-only memory – shared across the device, low latency, writable by CPU, readable by the computing device.
- local memory – smaller than global memory, shared by groups of processing elements (compute units) on the computing device.
- per-element memory – exclusive to each processing element.

The hierarchical OpenCL model of a device is shown in Figure 3.2.

Since each accelerator device uses its own memory, data have to be copied to the device for processing and the results have to be copied back from the device. This adds latency to the overall process, since moving data is magnitudes slower than performing computations on these data [41]. In addition, accelerator devices are specialized for crunching numbers in a fast and efficient way. Components are specially designed for this purpose. However, it comes at the cost of execution speed of more complex functions, jumps and unpredictable methods in general (e.g. recursion). Loops for example increase overall execution time of processing kernels. Therefore, these should be avoided by using work packages, unrolling or vector datatypes instead.

Work packages are multiple, parallel running instances of the same execution kernel. In order to achieve the high parallelism on the accelerator devices, the work carried out by the device has to be split up into work packages. This can be done by specifying work dimensions as well as work elements for each dimension. Within the kernel executed on the accelerator, the current dimension and element id can be retrieved and the appropriate data can be selected. In addition, OpenCL introduces its own set of datatypes focused on parallelization. Based on the well known datatypes like integer or double, OpenCL introduces vectorized versions of them. Supported vector sizes are 2, 3, 4, 8 and 16. In addition to the common sizes of 2^n, 3 is introduced to describe the common case of a vector containing three-dimensional coordinates.

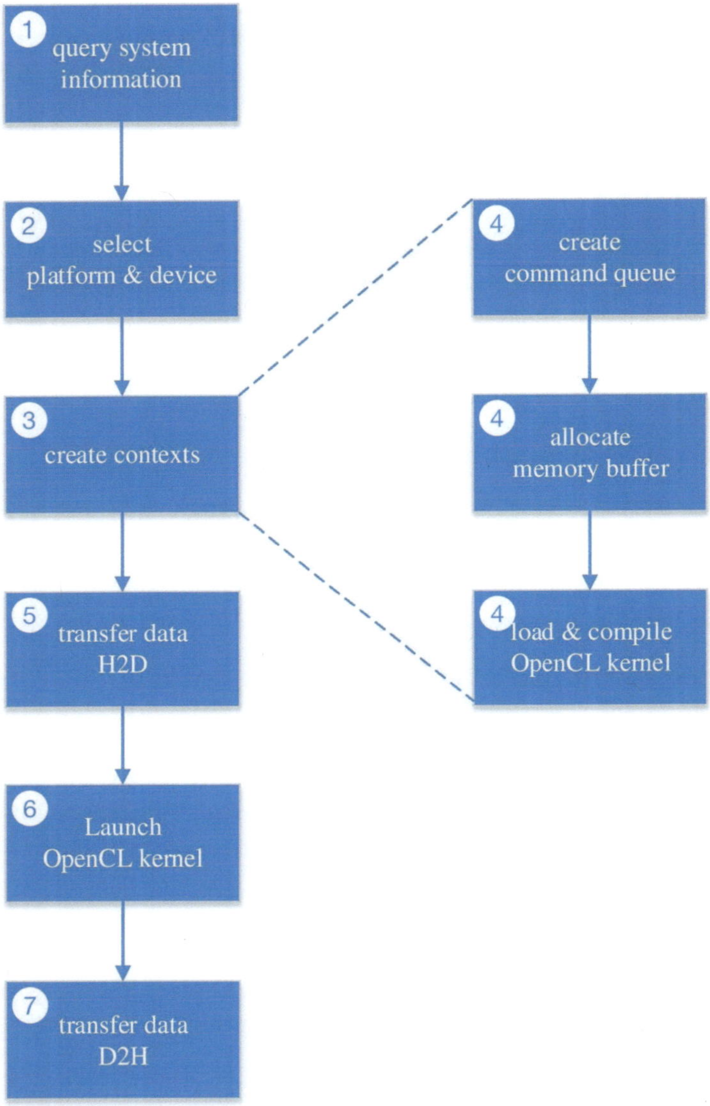

Figure 3.2: Program flow of an OpenCL application [8]. First, the system is queried for its capabilities and a platform and device are selected to run the program kernel on. Then, an execution context and a command queue are created and memory buffers are allocated. Data is transferred to the device and the OpenCL kernel is executed. Finally, data is transferred back from the device to the host.

The size is limited to 16 to allow access to each component by a single hexadecimal digit. These vector data types can then be used to carry out operations on each component in one step.

The individual steps necessary for running an OpenCL program can be seen in Figure 3.2.

1. The system has to be queried for available platforms and devices supporting the OpenCL standard.
2. One platform can have multiple devices attached.
3. A context for a certain platform is created.
4. Within this context, a command queue is created, memory buffers are allocated, and the OpenCL program code is loaded and compiled.
5. Once that is done, data can be transferred from the host to the device.
6. When the data is copied, the execution kernel can run.
7. After the kernel execution, the data from the device is copied back to the host.

OpenCL has been developed and is maintained by the Khronos Group [5].

3.7 Artificial Neural Networks

Artificial neural networks are an abstract representation of the human brain in computer science [48]. Neural networks are composed of so called neurons and connections between the neurons. A neuron calculates a sum of the input data passed to them. On this sum, an 'activation function' is applied. This function determines, whether the the neuron passes on the information it received. Commonly used activation functions are the Sigmoid function

$$S(x) = \frac{1}{1 + e^{-x}} \tag{3.1}$$

or the hyperbolic tangent

$$\tanh(x) = \frac{e^{x} - e^{-x}}{e^{x} + e^{-x}} \,. \tag{3.2}$$

Neural networks can be classified by the way the neurons are connected to each other and whether the information is passed unidirectional or bidirectional between the neurons.

Single Layer Feed Forward A network consists of only one layer, which is the output layer. Here, input neurons are directly connected to output neurons. Feed forward in this context means, that information are only passed from the input neurons to the output neurons.

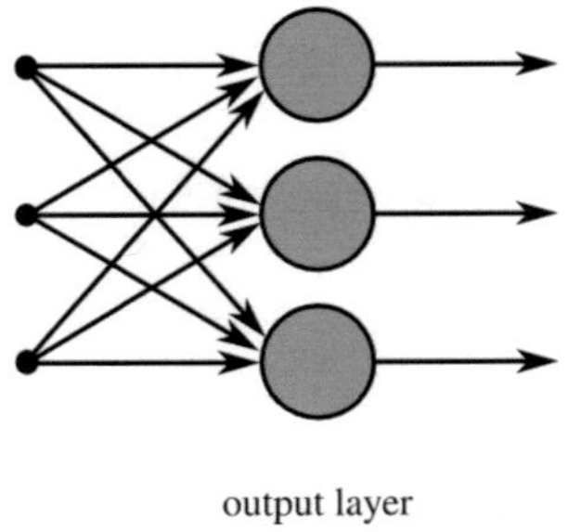

Figure 3.3: Example of a single layer feed forward network [6]. Each input neuron is directly connected to each output neuron.

Multi Layer Feed Forward Networks in this category have one or many hidden layers between the input and output layer. By utilizing hidden layers, a neural network is able to approximate more complex problems. One example of a more complex problem is the XOR-problem [66], which can not be solved using one layer of real-valued neurons.

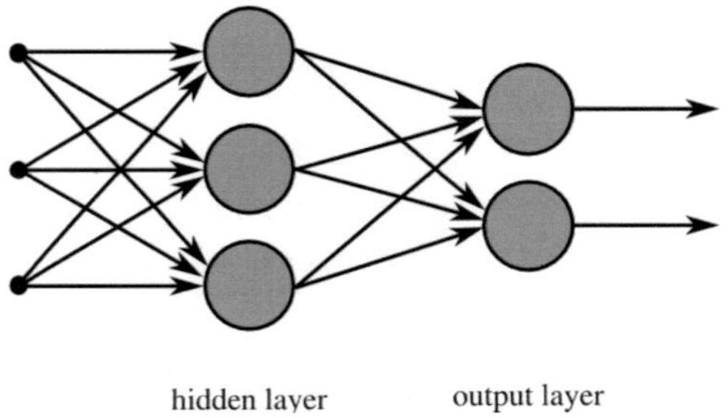

Figure 3.4: Example of a multi layer feed forward network with one hidden layer [7].
Here, each input neuron is connected to each hidden neuron. The hidden neurons are in turn connected to each output neuron.

Recurrent Network Recurrent neural networks have feedback loops allowing information to be passed into the same neuron again or back up to previous layers. This enables these types of networks to re-input earlier inputs by delaying the information passed through the feedback loops. It can be compared to memory, using previous calculations to affect later output. The recurrent connection is time delayed in order to incorporate output into later calculations.

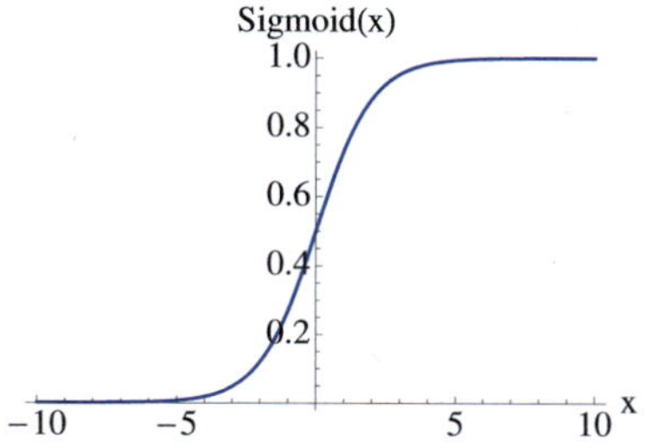

Figure 3.6: The Sigmoid function from $x = -10$ to $x = 10$.

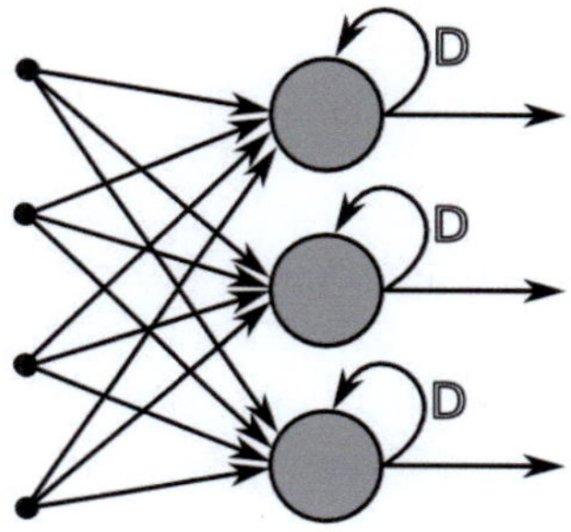

Figure 3.5: Example of a recurrent neural network [10]. In addition to the connections between input and output layer, the output neurons are also connected to themselves. Data is fed back to the same neuron after a time delay (D)

The connections between the neurons have weights attached. The weights influence the value of the signal passed on to the connected neuron. They can either increase, decrease or keep their original value.

For each neuron, the discrete inputs by all connected neurons are multiplied by the weights of the respective connection. Then their sum is calculated and used as the input for the activation function. The activation function itself acts as a threshold. The actual cut approximated by the activation function assigned to the neuron. The Sigmoid function for example (see Equation 3.1) returns 0.5 for $x = 0$, 1 for large positive inputs x and -1 for large negative of x, see Figure 3.6. The result of the activation function is the output of the neuron.

The weights of the connections between the neurons encode the knowledge about a problem. To solve problems, these weights have to be defined, usually through the process of training. The basis for training is a dataset for which the classification is known. This set should be as close to reality as possible to ensure that the network is able to solve variations of the same problem. The dataset is then split into three subsets: one for training (70 %) , one for validation

(20 %) and one for testing (10 %). The training and verification sets are used during training. The testing set is used after training is completed to verify the networks accuracy.

During training, the training set is processed and the difference between the known and the classification by the network is calculated. The weights of the network are then adjusted using a training algorithm. The validation set is used to ensure the network is not overly fitted to the training set. While processing the validation set, no weights are adjusted. This is repeated until the error of the network is sufficiently small. It should not be too small, because this could indicate an overtrained network, which may not be general enough to be applied to new data. Since the true classification must be provided beforehand, this form of training is called 'supervised training'. Many different algorithms exist for supervised training. One of the most famous is the backpropagation algorithm [50], see Section 3.7.1.

In addition to supervised training, a network can also be trained unsupervised or by reinforcement. In unsupervised training, only unrelated and unlabeled input data are provided. No feedback about the output is provided and the network is supposed to find a categorization on its own. The goal here is to study the changes of the neural network for different data provided. The outcome might be the discovery of new factors or labels common across previously unrelated data. The article 'Building High-level Features - Using Large Scale Unsupervised Learning' [58] for examples successfully explores the ability of face recognition by neural networks using unsupervised training.

Another training alternative is reinforcement training. Here, no training data are used at all. Instead, a task is given as an input for the network to solve. After it has been solved, feedback is given about the cost efficiency of the solution. The cost are usually determined by the influence of the solution to its surroundings. This process is repeated several times aiming for the most efficient solution, which can then be carried out. Reinforcement training is for example used in autonomous navigation. Since for this task it is not feasible to pre-train all possible routes and directions [70], they have to be calculated and weighed on the go. Here, a neural network can propose different routes and a cost function can then return a value based on speed limits, road quality and so on. Based on previous decisions, patterns can be formed, enabling the network to come to optimal decisions faster.

The input data for a neural network are commonly normalized to either $[-1, 1]$ or $[0, 1]$. This practice has several reasons. Firstly, input data may vary in their unit or order of magnitude, making it hard to put them into relation to each other [86]. Secondly, the likelihood of getting stuck in local minima may be reduced and a global minimum may be determined faster. An example for a feed forward neural network is shown in Figure 3.7.

3.7.1 Backpropagation

Backpropagation in the context of neural networks is a commonly used training algorithms for supervised learning and is short for 'backward propagation of errors' [91]. The goal of backpropagation is to determine a set of weights for a given neural network, that best maps an input vector to a desired output vector. The network itself can be seen as a complex mathematical function, which accepts numerical input and produces a numerical output. In the process of training, the weights of the connections are adjusted until the actual output is close to the ideal one.

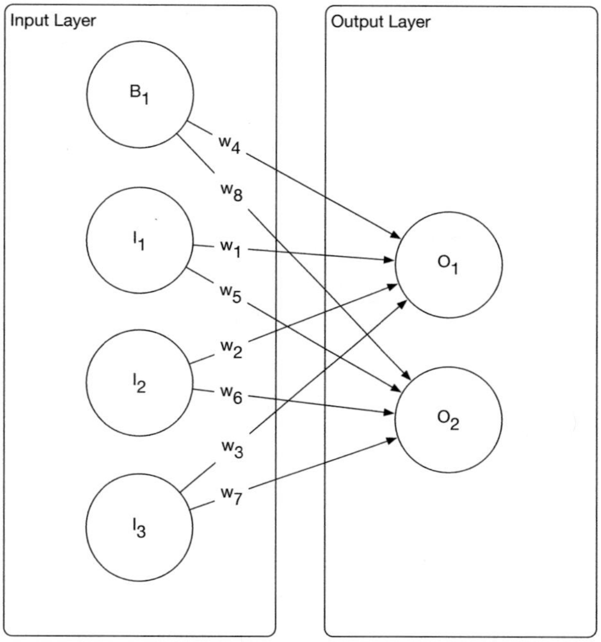

Figure 3.7: Example for a typical single layer feed forward neural network. The input layer consists of three input neurons (I_1, I_2, I_3), each one taking a discrete input and one bias neuron (B_1) which is set to 1. The output layer also consists of two neurons, O_1 and O_2.

The Backpropagation algorithm consists of two recurring phases. In the first phase, the training data are fed to the network and the output is calculated. This is done by gradually calculating the individual output values for each neuron. Taking Fig 3.7 as an example, the output for the neurons O_1 and O_2 would be calculated as

$$O_1 = \mathrm{S}(\mathrm{I}_1\mathrm{w}_1 + \mathrm{I}_2\mathrm{w}_2 + \mathrm{I}_3\mathrm{w}_3 + \mathrm{w}_4) \tag{3.3}$$

and

$$O_2 = \mathrm{S}(\mathrm{I}_1\mathrm{w}_5 + \mathrm{I}_2\mathrm{w}_6 + \mathrm{I}_3\mathrm{w}_7 + \mathrm{w}_8) \tag{3.4}$$

where S is the Sigmoid function (see Equation 3.1).

For each neuron, the ideal output $O_i^{(\text{ideal})}$ is then compared to the actual output using the quadratic error function

$$\text{Error}_1 = \frac{1}{2}\left(\text{O}_1^{(\text{ideal})} - \text{O}_1\right)^2 \tag{3.5}$$

$$\text{Error}_2 = \frac{1}{2}\left(\text{O}_2^{(\text{ideal})} - \text{O}_2\right)^2 . \tag{3.6}$$

The calculated error for all output neurons is then summed up

$$\text{Error}_{\text{total}} = \text{Error}_1 + \text{Error}_2 \tag{3.7}$$

returning the total error of the network.

In the second phase, the weights of the connections are updated to reduce the total error of the network. This is done by traversing the layers backwards.

Firstly, the new weights connected to the output neurons O_1 and O_2 are determined by calculating the impact of each connected weight on the error function. The new weight w_i' is calculated by

$$w_i' = w_i - \eta \frac{\partial \text{Error}_{\text{total}}}{\partial w_i} \tag{3.8}$$

where η is the learning rate. The higher the value of the learning rate, the faster the network is trained.

$\partial/\partial w_i$ denotes the partial derivative with respect to w_i.

The lower the value, the more accurate the training is, in general. However, the general problems are well-known, if the learning rate is too small or too large, respectively, the convergence may be slow or no minimum may be found as the algorithm may oscillate around a minimum.

By applying the chain rule we obtain

$$\frac{\partial \text{Error}_{\text{total}}}{\partial w_i} = \frac{\partial \text{Error}_{\text{total}}}{\partial O_k} \frac{\partial O_k}{\partial \text{In}_\text{j}} \frac{\partial \text{In}_\text{j}}{\partial w_i} \tag{3.9}$$

where Einstein's summation convention is adopted (implicit summation over repeated indices).

O_k is the output of the k-th neuron, see Equations (3.3) and (3.4), and

$$\mathrm{In}_1 = \mathrm{I}_1\ \mathrm{w}_1 + \mathrm{I}_2\ \mathrm{w2} + \mathrm{I}_3\ \mathrm{w}_3 + \mathrm{w}_4 \tag{3.10}$$

$$\mathrm{In}_2 = \mathrm{I}_1\ \mathrm{w}_5 + \mathrm{I}_2\ \mathrm{w6} + \mathrm{I}_3\ \mathrm{w}_7 + \mathrm{w}_8 \tag{3.11}$$

are weighted sums of the input neurons.

Furthermore

$$\frac{\partial \mathrm{Error}_{\mathrm{total}}}{\partial O_k} = -\left(O_k^{(\mathrm{ideal})} - O_k\right) \tag{3.12}$$

$$\frac{\partial O_k}{\partial \mathrm{In_j}} = O_k(1 - O_k)\ \delta_{kj} \tag{3.13}$$

where δ_{kj} is the Kronecker symbol, since

$$O_k = \frac{1}{1 + e^{-\mathrm{In_k}}} \tag{3.14}$$

Finally,

$$\frac{\partial \mathrm{In_j}}{\partial w_i} = I_i\ \delta_{ji} \tag{3.15}$$

provided the i-th weight is connected to the j-th input neuron.

Each w_n' is calculated. Then all weights of the neural network are updated. The algorithms is applied until the calculated error of the network is either sufficiently low or not improving anymore, or the maximum amount of iterations is reached.

Resilient Backpropagation

Resilient Backpropagation, or 'RPROP' is an improved version of the Backpropagation algorithm [27]. It introduces several optimizations compared to backpropagation.

Backpropagation uses the magnitude of the partial derivative to determine the change of weights. This is problematic in case of larger changes, since a narrow optimum might be missed. To mitigate this, a small learning rate is commonly chosen. However, this heavily increases the time necessary to train the network, since each iteration only slightly changes the network.

RPROP introduces two significant changes. Firstly, rather than using the calculated magnitude, only the sign of the gradient is used. Secondly, instead of a global learning rate, a dynamic learning rate for each weight is set and adapted during training. In each iteration, the previous and current signs are compared. If the sign is the same, the weight is updated by adding the current learning rate and the learning rate is increased. If the sign changes, the learning rate is decreased and the previous weight remains unchanged.

Usually, the main advantage of RPROP compared to the traditional backpropagation is a shorter training period for a given precission. The learning rate is determined dynamically [12], which reduces the chance to oscillate around narrow minima. One characteristic aspect of RPROP is that the learning rate is determined and adjusted dynamically per weight.

Variants In addition to the original resilient backpropagation algorithm three popular variations of the algorithm exist. For better distinction, they have been assigned different names by Igel et al. [52]

The variants are

- RPROP+ – The first version of RPROP
- RPROP- – RPROP without backtracking
- iRPROP- – Simplified version for easier implementation [76]
- iRPROP+ – Robust and typically faster than the above [75]

In our work the RPROP+ algorithm is used, since it is the default implementation in the Encog framework (see Section 3.10).

3.8 Block-Matching 3D

One example of a sophisticated noise removal algorithm is 'Block-Matching 3D' (BM3D) [37], which is a further improvement of the work presented in the article 'Image denoising with block-matching and 3D filtering' [38]. BM3D is considered as a very effective state of the art algorithm and works in multiple steps in order to remove noise from an image. The workflow of BM3D is illustrated in Figure 3.9.

The work-flow is divided into two separate steps. In step one

- similar sub-images are grouped by a block-matching algorithm using a reference image and a pre-defined distance for similar blocks,
- for each group, all images are transformed into the Fourier space,
- a hard cut is applied to separate signal from noise,
- each group is converted back using an inverse Fourier transformation,
- the groups are aggregated into one image,
- and estimates of smoothed images of each group are gathered for use in the next step.

Figure 3.8: Image processing by Block-Matching 3D. Left, an image with noise. In the center, the first step of BM3D has been applied, already removing most of the noise. On the right, the final version of the image is shown after both steps of BM3D are applied [59].

In step two the results are refined using a Wiener filter. Here,

- the already processed as well as the original groups of sub-images are used,
- both groups are independently transformed into the Fourier space,
- a Wiener filter is applied with the processed group as the desired signal and the original groups as the signal to be filtered,
- the reverse Fourier transformation is applied to the result of the Wiener filter,
- and the image is aggregated.

An example of a noisy image processed by BM3D can be seen in Figure 3.8. On the left, the source image is shown. In the middle the first step of BM3D is applied and on the right, both steps are applied.

It can be seen that most of the random noise is already removed after the first step. An open source implementation of the algorithm is proposed by Lebrun in the article 'An analysis and implementation of the BM3D image denoising method' [60].

3.9 Field Programmable Gate Array (FPGA)

Field-programmable gate arrays (FPGAs) are integrated circuits which can be reprogrammed after they have been manufactured [64]. FPGAs are programmed using a special 'Hardware Description Language'. Logic blocks can either perform simple Boolean operators like AND and OR, or can be configured to carry out combinational functions. Combinational functions are implemented using Boolean logic to express more complex functions. Some FPGAs also contain memory units for temporary storage. They can also contain analog parts, e.g., for signal processing.

Each FPGA is composed of programmable logic blocks that are connected. These blocks can be reprogrammed and reconnected. Since FPGA designs offer very fast I/O components and data buses, they can be programmed to efficiently manipulate and process data. Their main advantage are the millions of logic gates built into them. These enable a massively parallel data processing, given the algorithm can be implemented in a parallelizable way. One example is an FPGA programmed to calculate the fast Fourier transformation used in many areas of research. It has already been shown that the performance of fast Fourier transformation improves significantly by using FPGAs [88]. Another example is encryption such as AES [34] or hash algorithms such as SHA [13], increasing the performance of the encryption whilst freeing CPU resources.

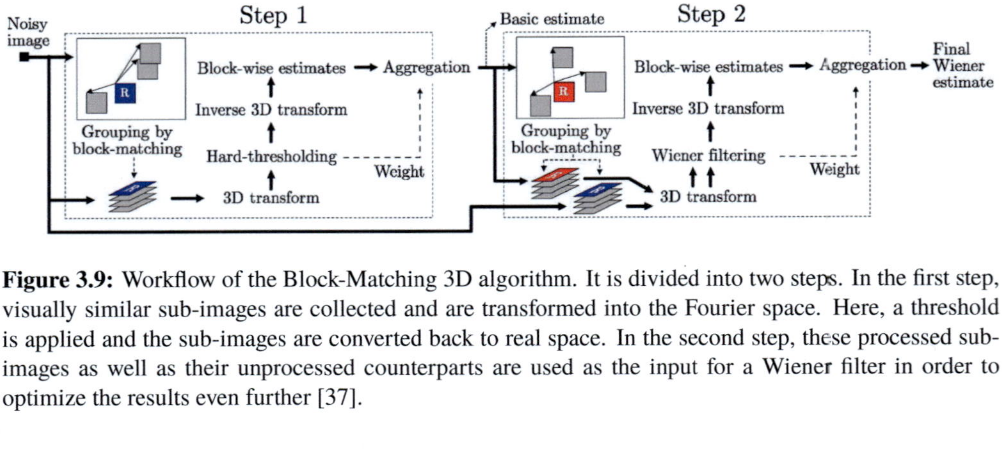

Figure 3.9: Workflow of the Block-Matching 3D algorithm. It is divided into two steps. In the first step, visually similar sub-images are collected and are transformed into the Fourier space. Here, a threshold is applied and the sub-images are converted back to real space. In the second step, these processed sub-images as well as their unprocessed counterparts are used as the input for a Wiener filter in order to optimize the results even further [37].

3.10 Encog Framework

Encog is a machine learning framework [49]. It is available in Java, C++, and .Net. All implementations support multithreading to utilize multicore machines. In addition, the C++ implementation of Encog supports OpenCL compatible GPUs (see Section 3.6) to offload parts of its computation.

Encog supports many different machine learning algorithms. Some examples are

- Support Vector Machines [36],
- Genetic Programming [16],
- Hidden Markov Models [42], and
- Artificial Neural Networks [48].

In addition, there are multiple training algorithms available for each learning algorithm. The framework also offers a GUI based 'workbench', which can be used to model and train machine learning algorithms.

Chapter 4
Problem Description

New generations of experiments in photon science such as the European XFEL [46] will be able to produce diffraction images at unpreceded volume and frequency. Given an image repetition rate of 27,000 images per second [15] and the resolution of the CSPad detector (see Section 2.2.1), roughly 400 TB of data will be generated per hour. When taking into account the small efficiency in photon science of 5 % [25] and less [62], approximately 380 TB of useless data would be stored every hour.

With these preconditions, it is neither practical nor desirable to store all data offline, in particular due to the sheer size of financial investments. Therefore, solutions have to be developed to pre-select the data online and as close to the detector as possible. In addition, a multi-level solution refining and possibly adding extracted and generated information during pre-selection would be desirable. This would enable indexing tools like CrystFEL(see Section 3.2) to use the coordinates of the Bragg spots found directly rather than extracting them again.

Since the amount of data is very high, possible solutions need to be parallelizable. It should also be explored, whether accelerator devices like Intel Xeon Phi [74] or General Purpose GPUs (GPGPUs) e.g. NVidia Tesla [61] are be able to increase the processing speed for individual images.

4.1 Test Data

In order to test and verify our approaches several runs of three different nanocrystallography experiments were provided by CFEL. Each run contains at least a few hundred images. Each experiment explored a different macro molecular structure.

The samples are

- the enzyme *Cathepsin B* (CatB) [72]),
- the *5-Hydroxytryptamine receptor 2B* (5HT-2B) [68],
- and the *granulovirus polyhedron* (GV) [33].

All images were exported from the Cheetah software and provided as individual files in the HDF5 format. In addition, files describing the geometry of the detector at experiment time were provided as well as a list of Bragg spots Cheetah found within the images along with their coordinates and intensities to compare our results to. The main difference between the samples is the intensity and distribution of random noise as well as the shape of the Bragg spots within an image. Background noise and shape of Bragg spots also vary within images of the same sample. Examples for the different shapes of Bragg spots can be seen in Figure 4.1. Figure 4.2 shows two examples of panels capturing multiple Bragg Spots. It can be seen that the shape as well as the relative intensity between the spots and the background varies.

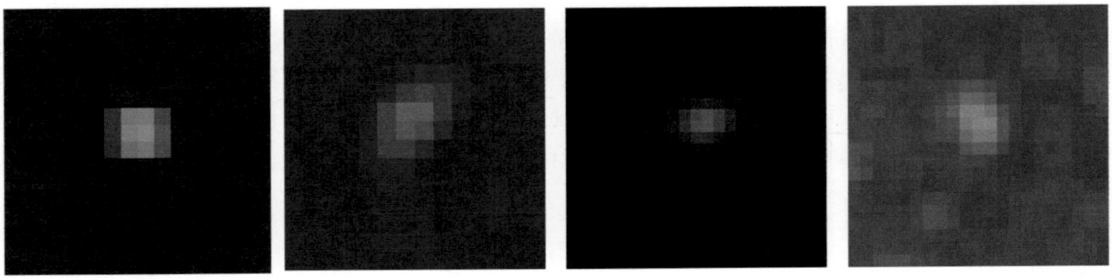

Figure 4.1: Examples for different shapes and intensities of signals. It can be seen that the intensities of the spots in the first and third images are stronger relative to the background vary. In addition, it can be seen that the spots can be pointlike but might also be sheared as in the second example.

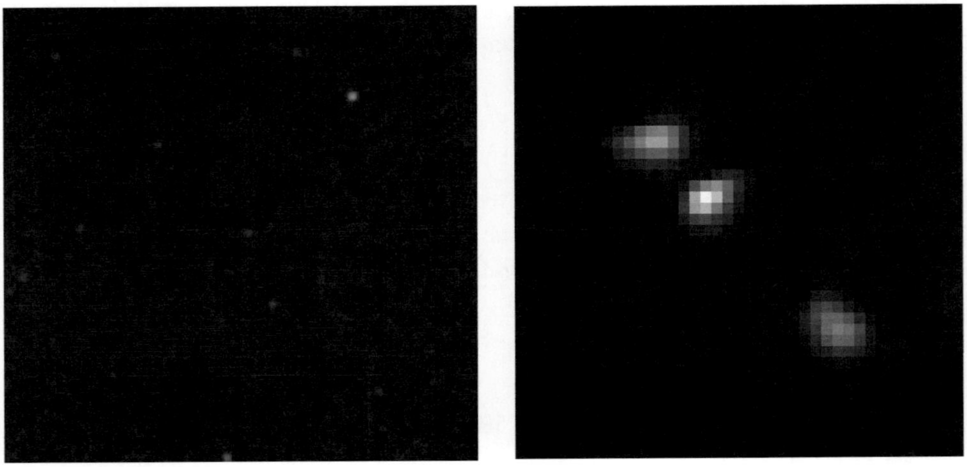

(a) Multiple Bragg spots.

(b) Three Bragg spots.

Figure 4.2: Image of one detector panel containing multiple Bragg spots of varying intensity (4.2a) and three Bragg spots of different shape and intensity (4.2b).

4.2 Data Verification

Cheetah (see Section 3.1) is currently used to identify Bragg spots. It is known that Cheetah is not able to identify all Bragg spots correctly. The article 'Serial Femtosecond Crystallography of G Protein-Coupled Receptors' [62] for example mentions a hit rate of $3.6\,\%$ according to Cheetah. Of these hits, Cheetah identified, only $21.5\,\%$ were indexed successfully by CrystFEL (see Section 3.2). However, there is no standard solution to compare spots found in an image to. The only way to ensure that an identified signal is a Bragg spot is manual verification by an expert. Also, indexing tools like CrystFEL (see Section 3.2) can be used to verify the Bragg spots found in an image. However, these tools merely test whether the possible Bragg spots allow reconstructing a molecule orientation. This does not imply the spots identified are valid and complete.

Since there is no reliable method besides an expert reviewing all Bragg spots found, we compare our results to Cheetah.

4.3 Data Normalization

Due to broken or stuck pixels of the detector, the data taken might contain values outside of the physically possible reading of $14\,\mathrm{bit}$ unsigned integer (16,364). Since these unwanted effects might interfere with our analysis, the intensities of all pixels outside these limits are set to zero. Let $I_i^{(raw)}$ be the intensity readout of the i-th pixel of an analyzed image. The quantity

$$I_i = \frac{I_i^{(raw)}}{16,364}\,\theta(I_i^{(raw)})\,\theta(16,384 - I_i^{(raw)}) \tag{4.1}$$

is called the intensity of the i-th pixel. $\theta(p)$ is the step function: it is zero for negative p, and one for non-negative p.

Data normalization is always applied before any further analysis is taking place.

4.4 Links to the research questions

If diffraction images were not affected by noise, a veto engine for selecting images with sufficiently many Bragg spots could easily be created. In practice, diffraction images show noise from different sources. An example of a blank image (no Bragg spots) containing only noise is shown in Figure 4.3. Besides random noise distributed throughout the image there is, in addition, a concentric ring of intense noise, the so called 'water halo' due to diffracted light from the transportation liquid. Around the center of the image, there is strong noise from the X-ray beam.

The software tool Cheetah, see Section 3.1, is used to search for Bragg spots offline. It is not designed to work under realtime conditions. Furthermore, its ability to recognize Bragg spots correctly is limited as already mentioned.

These considerations indicate the following approach. Firstly, a veto engine close to the detector should be developed which rejects all images which do not contain diffraction information.

This is reflected in research question '**1. How is it possible to determine if there is adequate data within an image at all?**'. Once only images that most likely contain data are passed on by the veto engine, the next question is, whether the data within the image are suitable for more sophisticated analysis. Here, the question is, whether it is possible to create an alternative to Cheetah, able to identify most of the Bragg spots within an image correctly while being able to process many images in parallel. This relates to research question '**2. Is the data within an image useful for further analysis?**'. Since noise and noise removal in images are a well known problem in image processing, the question arises whether and to which degree algorithms from image processing can be used to remove the noise beforehand. This is reflected in research question '**4. Can existing algorithms be used or adapted to facilitate the image optimization?**'. Given it is possible to develop solutions categorizing images and identifying signals in parallel, it is also important to meet the realtime constraints. These are currently dictated, as mentioned previously, by the European XFEL experiment taking 27,000 images per second. Therefore possible solutions need to be explored in terms of their runtime behavior and ability to scale up to this data rate. This is considered in research question '**3. Is it possible to solve the previous two questions within real-time constraints?**'.

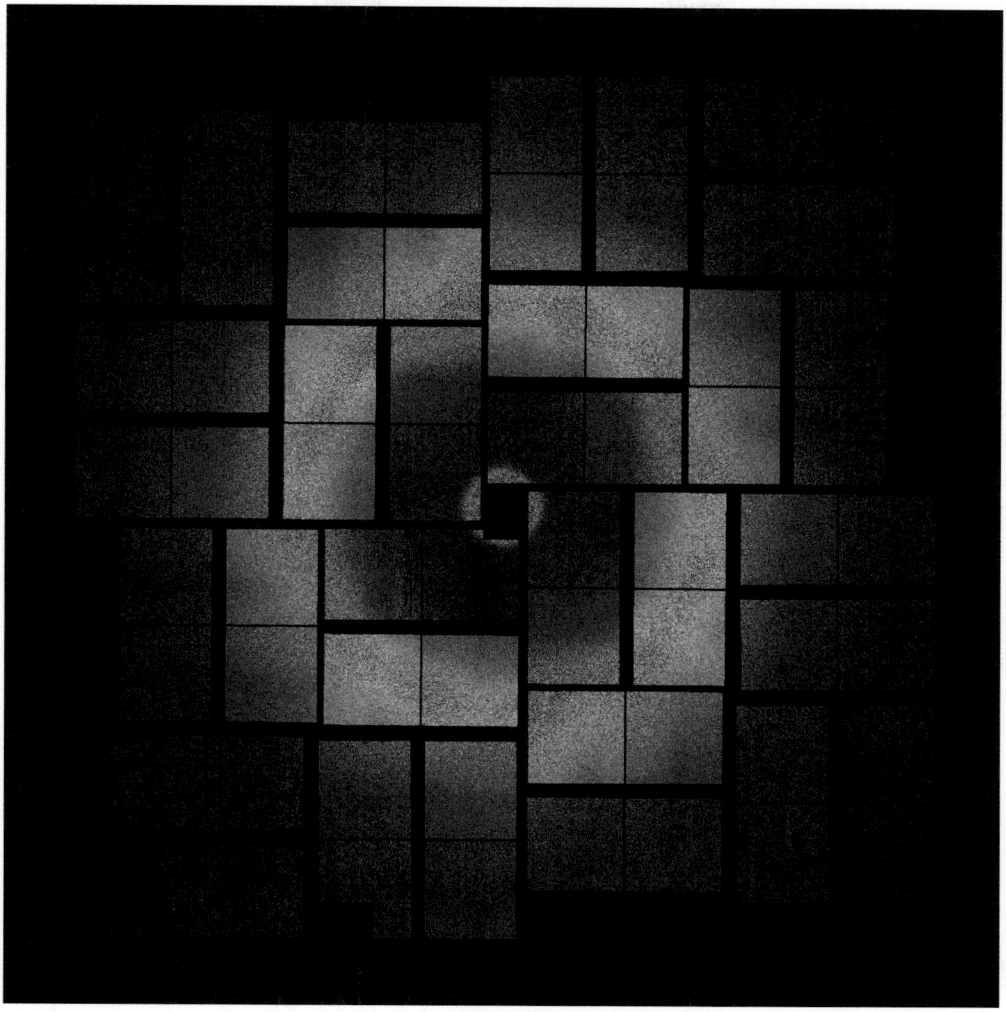

Figure 4.3: Example of an empty image with geometry applied. Random noise is distributed throughout the image. A concentric ring of intense noise (water halo) originating from diffracted light by the transportation liquid is visible. Strong noise around the center is caused by the X-ray beam.

Chapter 5

Neural Network as a veto engine

The following chapter discusses our research on neural networks used as a veto engine for data taken in nanocrystallography. Here, we are addressing the research question **1. How is it possible to determine if there are adequate data within an image at all?** The findings have been published in the article [18]. The content of the paper is covered in full in this chapter.

5.1 From complex to basic data

Large neural networks are successfully used to recognize handwriting. For example digits [35]. Here, all pixels of an image are used as an input vector. The benchmarked networks also contained multiple hidden layers.

At the beginning of our analysis we created a very large neural network, containing one input neuron for every pixel of an image. In addition, a hidden layer containing the same amount of neurons was used as well as two output neurons, indicating whether an image is suitable for further analysis or not. The resulting network consists of 2,296,660 input neurons, and the same amount of hidden neurons. In combination with the large number of input data, training and analysis took a long time because of the many calculations that had to be performed for each processing step of the network. In addition, the recognition rate was very poor.

The data showed that the established approach of using all pixels as an input is not feasible or might require even larger networks. Increasing the size and thus complexity of the network would result in even longer execution time for computations. Because of the realtime constraints, exploring even larger networks is not a feasible option. Therefore, a more thorough analysis of the data had to be done in order to fully understand the data we were trying to categorize.

The article 'Selection of radio pulsar candidates using artificial neural networks' [43] shows the utilization neural networks to identify pulsar candidates. The neural network is composed of 12 input and 12 hidden neurons, working on 12 features explicitly derived from data.

After a more thorough analysis of our data, we surprisingly found that three quantities extracted from an image are sufficient to achieve a high recognition rate.

We define an image as an ordered list of pixels

$$I = I_1, \ldots, I_n \tag{5.1}$$

where

$$n = lm \tag{5.2}$$

is the number of pixels and l, m are the height and width of an image, respectively.

The extracted quantities are

- the maximum intensity of all pixels of an image, I_{max},
- the average intensity of all pixels within an image, I_{mean},
- as well as the standard deviation of the average, ΔI.

The maximum intensity is defined as

$$I_{\text{max}} = \max_i (I_i) \tag{5.3}$$

, the mean intensity is defined as

$$I_{\text{mean}} = \frac{1}{n} \sum_{i=1}^{n} I_i \tag{5.4}$$

and the standard deviation as

$$\Delta I = \sqrt{\frac{1}{n} \sum_{i=1}^{n} (I_i - I_{\text{mean}})^2} \tag{5.5}$$

or

$$\Delta I = \sqrt{\frac{1}{n} \sum_{i=1}^{n} (I_i^2) - \left(\frac{1}{n} \sum_{i=1}^{n} I_i \right)^2} \,. \tag{5.6}$$

The second form is used later on to determine the standard deviation within one iteration over the image.

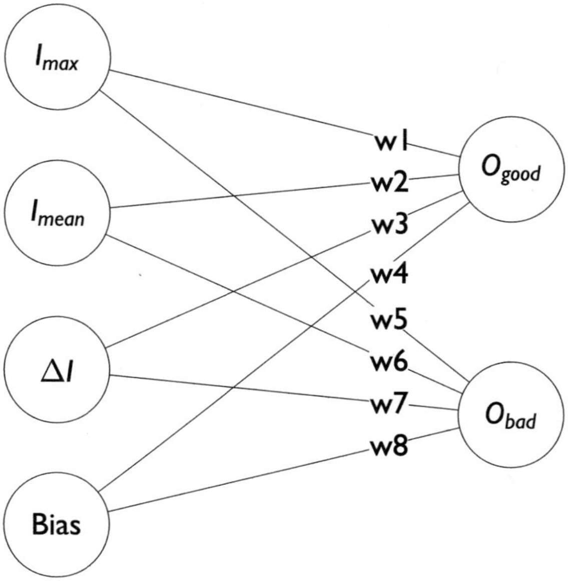

Figure 5.1: Layout of our neural network.

5.2 Architecture of the Neural Network

Based on the three quantities, a single layer feed forward neural network (see Section 3.7.1) has been developed using (I_{max}, I_{mean}, ΔI, 1) as the input vector. The fourth component results from the introduction of a bias neuron to the input. It is used to improve the approximation. Two output neurons are used, indicating whether an image is useful for further analysis or not. A hidden layer is not introduced. Resilient backpropagation is used for training (see Section 3.7.1). The layout of the network is shown in Figure 5.1.

5.3 Output Calculation

The output of the neuron O_{good} is used to identify indexable images and is given by

$$O_{good} = \mathrm{S}(\mathrm{w_1 I_{max}} + \mathrm{w_2 \Delta I} + \mathrm{w_3 I_{mean}} + \mathrm{w4}) \ . \tag{5.7}$$

Similarly, the output of the neuron indicating the likelihood of an image being non-indexable is given by

$$O_{bad} = \mathrm{S}(\mathrm{w_5 I_{max}} + \mathrm{w_6 \Delta I} + \mathrm{w_7 I_{mean}} + \mathrm{w8}) \tag{5.8}$$

after the network has been trained. An image is considered as indexable, if $O_{good} > O_{bad}$.

5.4 Data Optimizations

During the analysis of the data we found that the recognition rates varied for the three different experiments. The recognition rates for the 5HT-2B and GV samples were lower compared to CatB. After a visual inspection of the images, we found that the level of noise in the images taken from 5HT-2B and GV is much higher than CatB. To address this, two optimization techniques have been introduced, aiming either to reduce noise or enhance the signal within a given image as much as possible in order to increase the signal-to-noise ratio.

5.4.1 Background Subtraction

Background subtraction estimates a default level of random per-pixel noise. To correct for this estimated baseline, the average is subtracted from each image.

Single pixel background noise within images varies for each experimental run (each run contains at least a few hundred images each, see Section 4.1). In addition, it slightly varies during the same run, as well. Therefore, it is not feasible to use one pre-defined level of noise for the entire image. The contribution of each pixel has to be considered individually to remove as much noise as possible upfront.

The transportation liquid can change in different experimental runs and sometimes even within the same run [17]. This is reflected in changes in the background noise. Therefore, the average background should be re-determined in these circumstances by a new set of blank images from time to time. In our experiments we assumed a static background noise for the given data.

Prior to an analysis, $K = 500$ blank images were selected. Based on these images, an average noise level per pixel is determined as

$$I_i^{(\mathrm{noise})} = \frac{1}{K} \sum_{k=1}^{K} I_{k,i}^{(\mathrm{blank})} \tag{5.9}$$

where $I_{k,i}^{(\mathrm{blank})}$ stands for the intensity of the i-th pixel in the k-th blank image.

The noise reduced image $\widetilde{I}$ is defined as

$$\widetilde{I}_i = \left(I_i - I_i^{(noise)}\right) \, \theta \left(I_i - I_i^{(noise)}\right) \, . \tag{5.10}$$

Consequently, the basic data with background subtraction are defined as

$$\widetilde{I}_{\text{max}} = \max_i \left(\widetilde{I}_i\right) \tag{5.11}$$

$$\widetilde{I}_{\text{mean}} = \frac{1}{n}\sum_{i=1}^{n} \widetilde{I}_i \tag{5.12}$$

$$\Delta\widetilde{I} = \sqrt{\frac{1}{n}\sum_{i=1}^{n}\left(\widetilde{I}_i - \widetilde{I}_{\text{mean}}\right)^2}\,. \tag{5.13}$$

5.4.2 Transverse Intensity

In scattering experiments, a particle is usually more likely registered at smaller than at larger scattering angles, see e.g. the well-known Rutherford formula. The scattering angle is approximately equal to ratio of the transverse momentum over the initial momentum. Diffraction and scattering are closely related topics. Therefore, it can be expected that photons in nanocrystallography are more likely registered around the center of the CSPad detector than farther away in the outer areas.

We were able to show that similar effects apply to nanocrystallography as well. In order to mimic the role of the transverse momentum we introduce the 'transverse intensity'

$$I^{\text{T}} = \sum_{i=1}^{n} \widetilde{I}_i \sin\vartheta_i \tag{5.14}$$

where

$$\vartheta_i = \text{atan}\frac{\sqrt{{x_i}^2 + {y_i}^2}}{z} \tag{5.15}$$

is the scattering angle between the i-th pixel, the interaction point between sample and laser and the beam axis. x_i and y_i are the coordinates of the i-th pixel relative to the center of the detector surface and z is the distance between the interaction point of the laser and sample and the detector plane at the beam hole. In case of the LCLS experiment, the distance z is set to $68\,\text{mm}$ (see Figure 2.3), since only the front panel is used in current experiments.

The transverse intensity is defined correspondingly

$$I^T_{\max} = \max_i \left(\widetilde{I}_i \sin \vartheta_i \right) \tag{5.16}$$

$$I^T_{mean} = \frac{1}{n} \sum_{i=1}^{n} \widetilde{I}_i \tag{5.17}$$

$$\Delta I^T = \sqrt{\frac{1}{n} \sum_{i=1}^{n} \left(\widetilde{I}_i - I^T_{mean} \right)^2} . \tag{5.18}$$

5.5 Signal-to-noise ratio

The signal-to-noise ratio is defined as

$$\mathrm{SNR} = \frac{\mathrm{P_{signal}}}{\mathrm{P_{noise}}} \tag{5.19}$$

where P is the average power of signal and noise, respectively. However, in our case the values for $\mathrm{P_{signal}}$ and $\mathrm{P_{noise}}$ are not known. The smaller $\mathrm{P_{noise}}$ compared to $\mathrm{P_{signal}}$, the larger is the quantity SNR, i.e. high values are an indicator of clear signals. Therefore, we define signal-to-noise ratio by

$$\mathrm{SNR} = \frac{\mathrm{I_{mean}}}{\Delta \mathrm{I}} \tag{5.20}$$

where I_{mean} is the mean intensity of an image and ΔI is the associated standard deviation of the image I. The smaller the standard deviation, i.e. the noise, the larger the value for SNR, as expected intuitively.

To calculate the average signal-to-noise ratio for K images, we define

$$\mathrm{SNR^{average}} = \frac{1}{\mathrm{K}} \sum_{\mathrm{n}=1}^{\mathrm{K}} \frac{\mathrm{I_k^{mean}}}{\Delta \mathrm{I_k}} \tag{5.21}$$

as the average signal-to-noise ratio.

5.6 Experimental setup

In order to test the proposed neural network, it has been implemented using the ‘Encog Machine Learning Framework’ [49] (see Section 3.10) and is written in Java. The architecture can be seen in Figure 5.1. The value of the bias neuron is set to 1 and the Sigmoid function is used as the activation function of the neurons. Due the smoothness of the function, it prevents individual neuron output from overpowering the network [78]. As previously mentioned, no hidden layer is used. The initial weights of the connections are randomized for each run. Resilient backpropagation is used as the training algorithm [27] (see Section 3.7.1) to determine the weights $w_1 \dots w_8$ without additional options. It is an improved version of the iterative backpropagation algorithm.

5.7 Results

To verify our results, the neural network has been trained and tested for all three different macromolecules (see Section 4.1). From each of the three molecules, 200 indexable and 200 non-indexable images according to Cheetah were randomly selected. The term 'indexable' means an image is suitable for analysis and 'non-indexable' means the images does not contain sufficient Bragg spots. 100 images were used for training and the other half was used for verification.

In Table 5.2 the calculated weights for the neural network are shown. The weights represent the impact of the input data on the rating by the output neurons. The weights and input neurons are connected as follows, see Figure 5.1:

- $I_{\text{max}} - w_1, w_5$
- $I_{\text{mean}} - w_2, w_6$
- $\Delta I - w_3, w_7$

The weights w_4 and w_8 are associated to the bias neuron. In the case of CatB, I_{max} has the largest impact, because its associated weights are the largest.

For the other two samples however, I_{mean} and ΔI have the by far greatest influence on the output of the network. This can be explained by the noise in the images. In the case of CatB, the noise level is rather low and most signals stand our clearly against the background. Therefore, the maximum intensity is a meaningful value. The images of the other two samples contain much more noise. In addition, the signals in the images of GV are very weak. This means that the maximum intensity is no clear indicator in these cases. Here, especially the standard deviation seems to be the most meaningful quantity.

Table 5.4 shows the recognition rates of the network for the different samples and optimizations. The previously discussed optimizations affect CatB in a positive way, further increasing the recognition rate. In case of 5HT-2B and GV, the recognition rate is affected negatively by the optimizations. This is probably due to the weaker signals in combination with more noise in the images. Most likely, by applying background subtraction, valid signals are dampened as well and transverse intensity also increases remaining noise.

The poorer recognition rates for 5HT-2B and GV can be explained by looking at the signal-to-noise ratio. Table 5.3 shows the calculated average signal-to-noise ratio using Equation 5.5 for a separate set of 50 indexable and 50 non-indexable images. In case of CatB, the average for signal-to-noise ratio of indexable is lower than for non-indexable images, whereas this is not the case for 5HT-2B and GV. An equally high signal-to-noise ratio for indexable and non-indexable images means that signals do not stand out against the background noise at all.

		w1	w2	w3	w4
CatB	***w/o optimization***	7.21	−3.96	−3.82	−1.9
	average subtraction	6.28	−0.26	−0.3	−1.58
	average subtraction & transverse intensity	7.14	−0.26	−0.17	−1.32
5HT-2B	***w/o optimization***	0.45	−13,357	10,553.77	−1
	average subtraction	0.97	−5082.37	2814.12	−2.44
	average subtraction & transverse intensity	0.97	−5082.37	2814.12	−2.44
GV	***w/o optimization***	3.15	−25,719.88	22,964.67	−3.09
	average subtraction	2.26	−3177.03	3176.71	−5.74
	average subtraction & transverse intensity	2.2	−2107.5	2097.33	−3.67

Table 5.1: Calculated weights w_1 to w_4 of the trained neural network.

		w5	w6	w7	w8
CatB	***w/o optimization***	−6.33	4.27	2.92	1.71
	average subtraction	−5.85	1.71	0.58	1.4
	average subtraction & transverse intensity	−6.25	0.94	−1.12	1.12
5HT-2B	***w/o optimization***	−0.45	13,351.74	−10,549.5	0.99
	average subtraction	−0.97	5082.34	−2814.1	2.44
	average subtraction & transverse intensity	−0.97	5082.34	−2814.1	2.44
GV	***w/o optimization***	−3.19	26,078.63	−23,283.5	3.11
	average subtraction	−2.26	3176.98	−3176.66	5.74
	average subtraction & transverse intensity	-2.2	2107.43	−2097.26	3.67

Table 5.2: Calculated weights w_5 to w_8 of the trained neural network.

	SNR indexable images	SNR non-indexable images	Δ
CatB	1.11	0.82	0.29
5HT-2B	0.75	0.74	0.01
GV	0.56	0.54	0.02

Table 5.3: Signal-to-noise ratio calculated using equation 5.21. The mean signal-to-noise ratio has been calculated for 50 indexable and 50 non-indexable images for each sample. In addition, the distance between the signal-to-noise ratio of indexable and non-indexable images is shown.

	w/o optimization	background subtraction	background subtraction & transverse intensity
CatB	88 %	90 %	93 %
5HT-2B	63 %	62 %	61 %
GV	79 %	74 %	70 %

Table 5.4: Average recognition rate of the neural network. True positive and true negative combined for each sample. Absolute recognition rates are shown without optimizations, with background subtraction applied as well as background subtraction and transverse intensity combined.

5.8 Summary

In this chapter we showed that three basic data extracted from an image are sufficient to categorize up to 93% of the images correctly compared to Cheetah. In order to perform the actual categorization, we introduced a simplistic neural network with only three input values and no hidden layer. We found that the main challenge in categorizing the data correctly is noise within the images. Categorization can be done reliably, given the signal-to-noise ratio is sufficiently high. To improve the recognition rates, two optimization techniques have been introduced. Firstly the background subtraction, which calculates an average noise level for each individual pixel, given a series of blank images. And, secondly, the transverse intensity, which we introduced as a new quantity in photon science.

Chapter 6
Signal Identification

In this chapter, it is shown why it is not possible to use most state of the art noise removal algorithms in nanocrystallography. Our findings have been published in the article [20] and are fully covered in this chapter. In addition, a combination of algorithms is introduced to remove the noise specific to nanocrystallography images, enhance signals, and detect signals within images. This research has been published in the articles [19] and [21]. This chapter addresses the research question **2. Is the data within an image useful for further analysis?** and **4. Can existing algorithms be used or adapted to facilitate the image optimization?**

Once an image has been found to contain data, the next step is to detect individual signals within the image. The article [17] states that based on practical experience, an image should contain at least 20 Bragg spots in order to be suitable for indexing using current tools like CrystFEL (see Section 3.2).

We found the main challenge in detecting diffraction information within an image to be random noise spread throughout the image. Therefore, it is desirable to remove as much noise as possible before searching for signals within an image. In the last Chapter, we introduced background subtraction as a basic tool for noise reduction and transverse intensity for the enhancement of signals at the outer areas of the detector. In this Chapter, these ideas are revisited by using convolution to dampen noise even further and enhance Bragg spots as well as unifying their shape to improve recognition. We also focus on the identification of individual signals within an image rather than classifying an image.

There are plenty of noise removal algorithms known in image processing. Sophisticated algorithms are able to achieve very good results for typical photographs. They rely on significant differences between the characteristics of the information a user wants to keep and the noise that should be removed. Figure 6.1a shows an original picture without noise. In Figure 6.1b 20 % random per pixel noise is added. Structures such as grassland, sky, and trees are still recognizable. The size of the objects is much larger than the noise spots. In nanocrystallography, however, noise and signal are both spot-like and cannot be separated that easily.

(a) Picture without noise

(b) Picture with 20% random noise

Figure 6.1: Example of a picture before and after adding 20 % random noise.

Figure 6.2: A typical image taken from the 5HT-2B sample (50x50 pixel). The small bright spots are the Bragg spots.

6.1 Separation of noise and signal in nanocrystallography

An example for signals in nanocrystallography is shown in Figure 6.2. The bright spots are the actual signals. As can be seen, they are only a few pixels in size. Dark areas or isolated low intensity pixels are background noise. In contrast to elements in photographies, the distinction between content and noise is much less pronounced. Bragg spots may have the same size as noise, namely individual pixels and vary in shape and size. Consequently, it is not feasible to use most of the algorithms in photography for removing noise from images in photon science.

This can be understood as a consequence of the uncertainty relation, according to which the resolution power of a Fourier series expansion is limited if the series is replaced by a finite sum. In other words, a function and its Fourier transform cannot both have a finite support. This is due to the fundamental Paley-Wiener theorem which is proven in textbooks on the theory of Fourier transformations, see e.g. [73].

It is instructive to illustrate the essential point of the theorem and the situation in photon science by simple examples. Let us consider a one-dimensional signal of Gaussian shape on the interval $0 \leq x \leq 2\pi$

$$\mathrm{s}(x) = s_0 \, e^{\frac{-(x-x_0)^2}{\sigma^2}} \tag{6.1}$$

where s_0 is the amplitude, x_0 the center and σ the width of the signal, see Figure 6.3. The noise is identified with a periodic signal

$$\mathrm{n}(x) = n_0 \, \sin(kx) \tag{6.2}$$

where n_0 is the amplitude and k the wave number.

The resulting additive noise signal

$$\mathrm{sn(x) = s(x) + n(x)} \tag{6.3}$$

, see the yellow curve in Figure 6.3, can be expanded into a Fourier series

$$\mathrm{sn(x)} = \sum_{\mathrm{m}=-\infty}^{\infty} \mathrm{f_m e^{imx}} \tag{6.4}$$

where the Fourier coefficients are given by

$$f_m = \frac{1}{2\pi} \int_0^{2\pi} \mathrm{sn(x) e^{-imx} dx} \tag{6.5}$$

The integral can be calculated explicitly and is given by

$$f_m = \frac{\sigma}{4\sqrt{\pi}} \left(\mathrm{erf}\left(\frac{\mathrm{x_0}}{\sigma} + \mathrm{ik}\frac{\sigma}{2}\right) + \mathrm{erf}\left(\frac{\mathrm{x_0}}{\sigma} - \mathrm{ik}\frac{\sigma}{2}\right) \mathrm{e}^{-\mathrm{k}(\mathrm{k}\frac{\sigma^2}{4} + \mathrm{ix_0})} \right) \tag{6.6}$$

where

$$\mathrm{erf(x)} = \frac{2}{\sqrt{\pi}} \int_0^{\mathrm{x}} \mathrm{e^{-t^2} dt} \tag{6.7}$$

is the error function.

The Fourier coefficients are shown in Figure 6.4. The two non-vanishing values of the imaginary part at the wave numbers m = -10 and 10 correspond exactly to the Fourier coefficients of the noise.

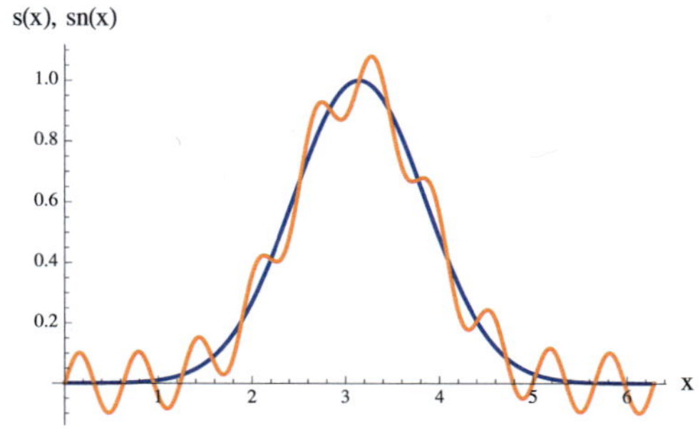

Figure 6.3: Blue curve: Gaussian signal around $x = \pi$, of the width $\sigma = 1$, Yellow curve: signal with added noise ($n_0 = \frac{1}{10}$, $k = 10$).

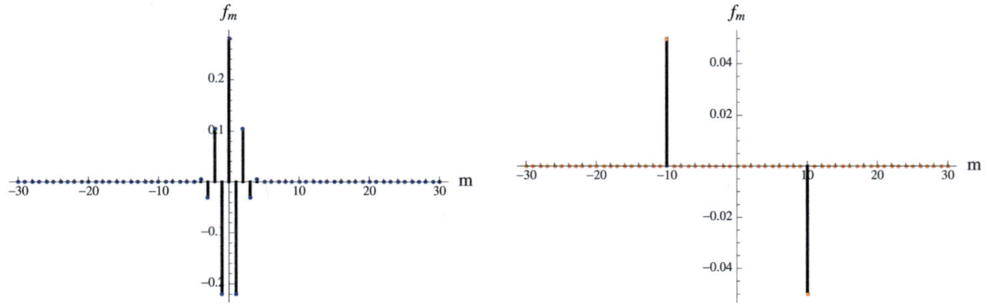

(a) Real part

(b) Imaginary part

Figure 6.4: Fourier coefficients of the signal s and additive noise n.

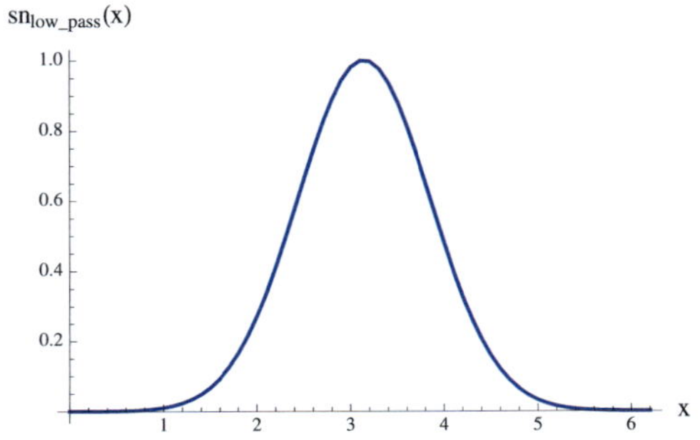

Figure 6.5: Reconstructed signal after application of a low pass filter. There is almost no difference between $\mathrm{sn}_{\mathrm{lowpass}}$ and the original signal. The influence of the noise is mostly removed.

Let us define a low-pass filter by

$$\mathrm{sn}_{\mathrm{low-pass}}(\mathrm{x}) = \sum_{\mathrm{m}=-5}^{5} \mathrm{f}_{\mathrm{m}} \mathrm{e}^{\mathrm{imx}} \tag{6.8}$$

i.e. only the wave numbers between -5 and 5 are taken into account.

Functions fluctuating strongly show large Fourier coefficients at large wave numbers, i.e. a function can be smoothed by applying a low-pass filter. Figure 6.5 shows that noise can be removed if the signal is sufficiently broad in relation to the noise. This corresponds to the typical situation in photography and many noise reduction algorithms are developed for this scenario.

However, in nanocrystallography both signal and noise are pointlike. Let us consider a narrow signal

$$s'(x) = s_0 e^{\frac{-(x-x_0)^2}{\sigma'^2}} \tag{6.9}$$

where $\sigma' \ll \sigma$, see Figure 6.6 where the width of the signal is comparable to the inverse of the noise's wavenumber, $\sigma' \approx \frac{1}{k}$.

The plot of the combined signal and noise can be seen in Figure 6.6, where the blue curve represents the function $s(x)$ and the modified function

$$\mathrm{sn}'(\mathrm{x}) = \mathrm{s}' + \mathrm{n}(\mathrm{x}) \tag{6.10}$$

is plotted in yellow.

The Fourier coefficients for the combined functions can be found in Figure 6.7. The narrow signal is considerably influenced by Fourier coefficients at wave numbers m beyond the Fourier coefficients at $|m| = 10$. A separation between noise and signal by applying a threshold is not possible, see Figure 6.8 showing the reconstructed signal after the same low-pass filter has been applied. The reconstructed signal is wider than the original one and, in addition, not all noise has been removed.

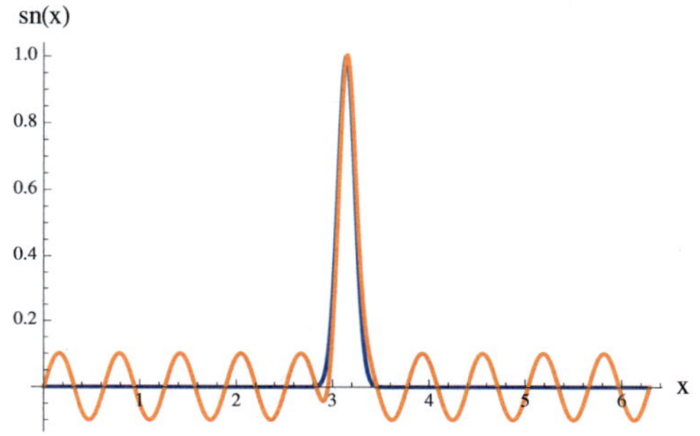

Figure 6.6: Blue curve: signal, yellow curve: signal plus noise ($s_0 = 1$, $x_0 = \pi$, $\sigma' = \frac{1}{8}$, $k = 10$).

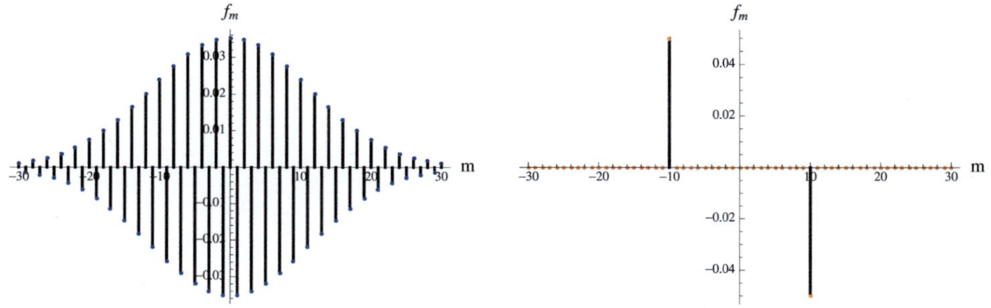

(a) Real part

(b) Imaginary part

Figure 6.7: Fourier coefficients of the narrow signal $s'(x)$ with noise.

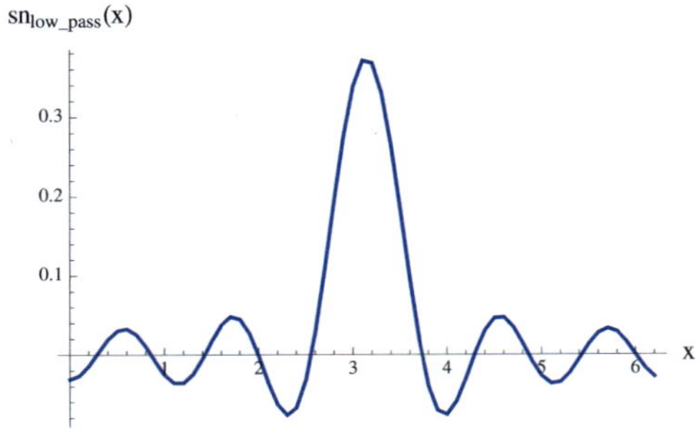

Figure 6.8: Reconstructed signal after the application of the low-pass filter $sn_{low-pass}(x)$.

This example illustrates why standard approaches for noise removal are not applicable in nanocrystallography. Thresholds in the Fourier space cannot be applied as thereby signals are removed as well. However, this technique works well for noise reduction in photographs and is used by modern noise removal algorithms like Block-Matching 3D (see Section 3.8).

6.2 Clusterfinder

As explained in the previous section, it is not feasible to use most state of the art algorithms for noise removal. However, this does not mean that basic algorithms from image processing cannot be adapted.

The clusterfinder algorithm uses established techniques of convolution [67] and edge detection for identifying signals in images.

It is composed of the steps

1. Reduction of single pixel noise
2. Edge detection
3. Signal identification

and will be discussed in detail in the following sections.

An image of the 5HT-2B sample is used to illustrate the optimizations introduced by each step. This sample had the worst recognition rate by the neural network discussed in the previous Chapter. This was mainly due to the amount of noise in the images. It is used as an example here to illustrate the efficient elimination of noise by the algorithms discussed in this Chapter.

A version without any optimizations (except data normalization, see Section 4.3) and with geometry applied (see Section 2.2.2) is shown in Figure 6.9. To help the eye, the contrast and brightness have been increased. Around the center, some of the light of the X-ray flash has been captured. The concentric dark circle mostly consists of diffracted light from the transportation liquid (water halo). Between the water halo and the center of the image, small black spots can be seen. These are Bragg spots.

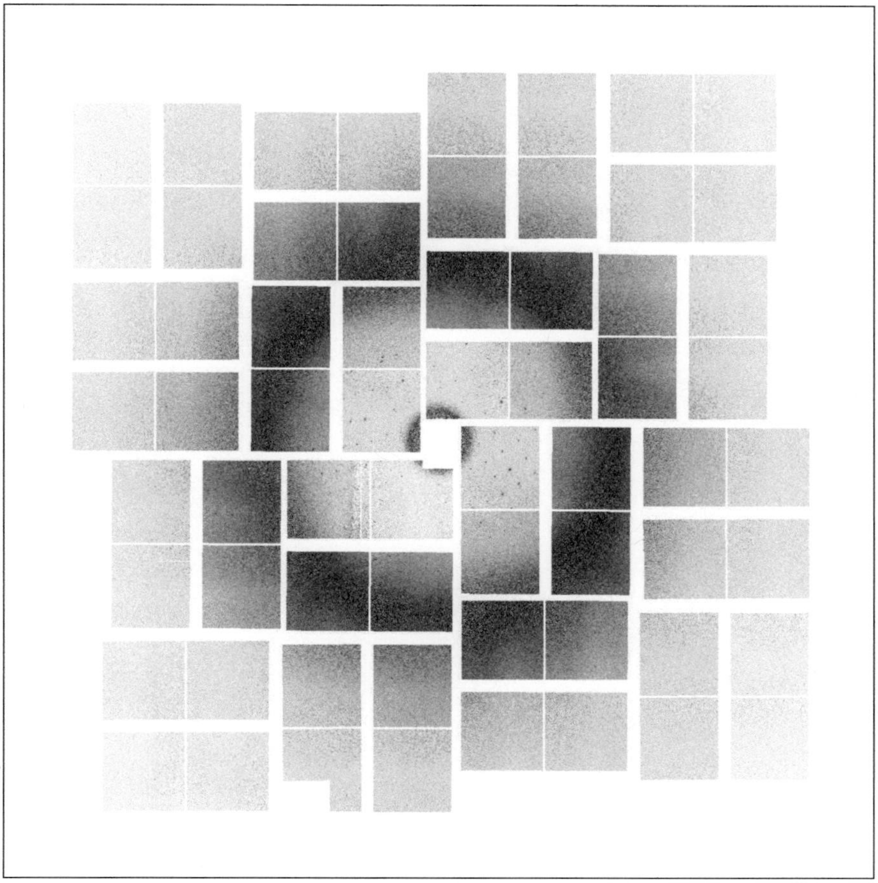

Figure 6.9: Image taken from the 5HT-2B sample without any optimizations applied (except for normalization, see Section 4.3). For better visualization, the contrast and brightness are increased, the intensity is inverted, and the panels are arranged in the correct physical geometry. At the center part of the laser beam can be seen as well as the beam hole. The concentric dark circle is due to light diffracted from the transportation liquid (water halo). Between the water halo and the center, small black spots (Bragg spots) are visible.

6.2.1 Reduction of single pixel noise

Convolution is used in image processing for manipulating images in the small. A small matrix (a so called image kernel, typically 3×3 or 5×5) is is applied to the area around the pixel. Using a symmetric kernel it is possible to blur or sharpen images, for example. With a non-symmetric kernel, edges within an image can be detected or the image can be obfuscated.

The convolution operator $*$ of the image I with the kernel K (of dimension $M' \times N'$)

$$I' = I * K \tag{6.11}$$

is defined via

$$I'_{x,y} = \sum_{m=1}^{M'} \sum_{n=1}^{N'} I_{x-m+2,y-n+2} K_{m,n}, \tag{6.12}$$

The image I is defined as a $M \times N$ matrix. In our case, the dimension of the image matrix I is either 1552×1480 (without geometry applied, see Section 2.2.2) or 2000×2000 with geometry applied.

There are several ways to deal with pixels at the border of images. In our case, the border area of an image is very unlikely to contain any useful information. Therefore, the limits $1 < x < M$ and $1 < y < N$ are applied.

Most of the random noise within the images is single pixel noise. In order to dampen this kind of noise, the image kernel

$$K = \frac{1}{9} \begin{pmatrix} 1 & 1 & 1 \\ 1 & 1 & 1 \\ 1 & 1 & 1 \end{pmatrix}. \tag{6.13}$$

is used.

Since K is symmetric and the sum of its components is 1, no intensity is lost and the intensity of each pixel is smeared across its neighbors, thereby dampening bright single pixels. The prefactor of $\frac{1}{9}$ is chosen to ensure that the total intensity of an image is not changed.

In Figure 6.10 single pixel noise reduction has been applied. Almost all random noise has been removed. Only the previously very noisy area around the beam hole at the center still contains some noise which is due to the strength of noise in that area. The Bragg spots now clearly stand out against the background.

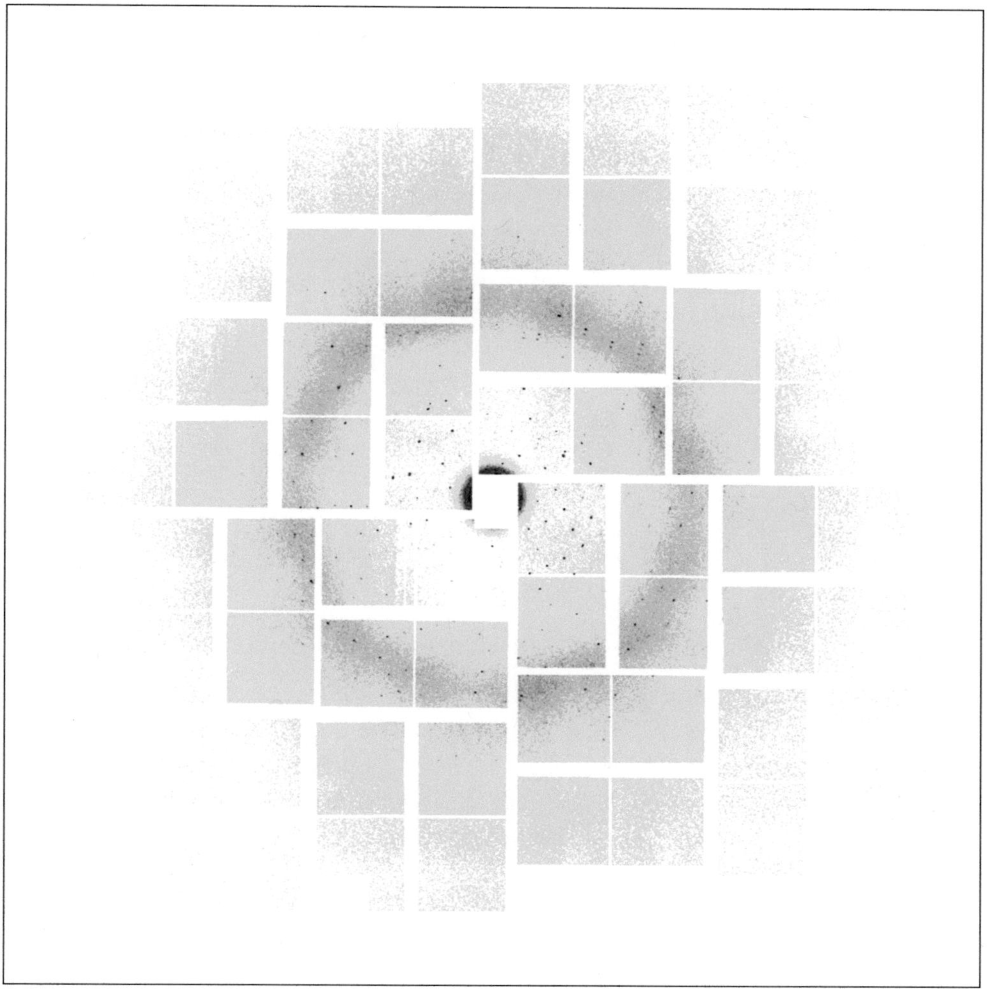

Figure 6.10: Sample image with noise reduction applied. It can be seen that most of the random noise throughout the image has been dampened. In addition, the water halo and noise around the center are dampened as well. The dark black Bragg spots now clearly stand out against the background.

6.2.2 Edge Detection

A challenge in identifying signals in images from nanocrystallography is their variation in shape and size (see Section 4.1).

However, what the spots have in common is the sudden increase in intensity from one pixel to another, which stays high for a couple of pixels and then suddenly drops down again. This characteristic can be exploited to extract the Bragg spots by applying edge detection.

Edge detection is a process in which the image is convoluted using multiple separate, non-symmetric kernels. The well-known Sobel operator [85] defined as

$$S_x = \begin{pmatrix} -1 & 0 & 1 \\ -2 & 0 & 2 \\ -1 & 0 & 1 \end{pmatrix} \text{ and } S_y = \begin{pmatrix} -1 & -2 & -1 \\ 0 & 0 & 0 \\ 1 & 2 & 1 \end{pmatrix} \tag{6.14}$$

is used to identify intensity jumps in the x- and y-direction, respectively. We use the quantity

$$\Delta I_{x,y} = \sqrt{(I' * S_h)^2_{\mathrm{x,y}} + (I' * S_v)^2_{\mathrm{x,y}}} \tag{6.15}$$

as a measure for the strength of the change in intensity at the pixel coordinates x, y.

Figure 6.11 shows the previously noise reduced image with edge detection applied. It can be seen that the black Bragg spots now stand out much more clearly. Since edge detection has been applied to the whole image, panel edges are highlighted as well, since all pixels between panels are 0 which represents a severe change in intensity due to the ambient noise of the panels. However, since the coordinates of the panel edges are known, the respective coordinates can either be ignored or removed (see Section 6.3.3).

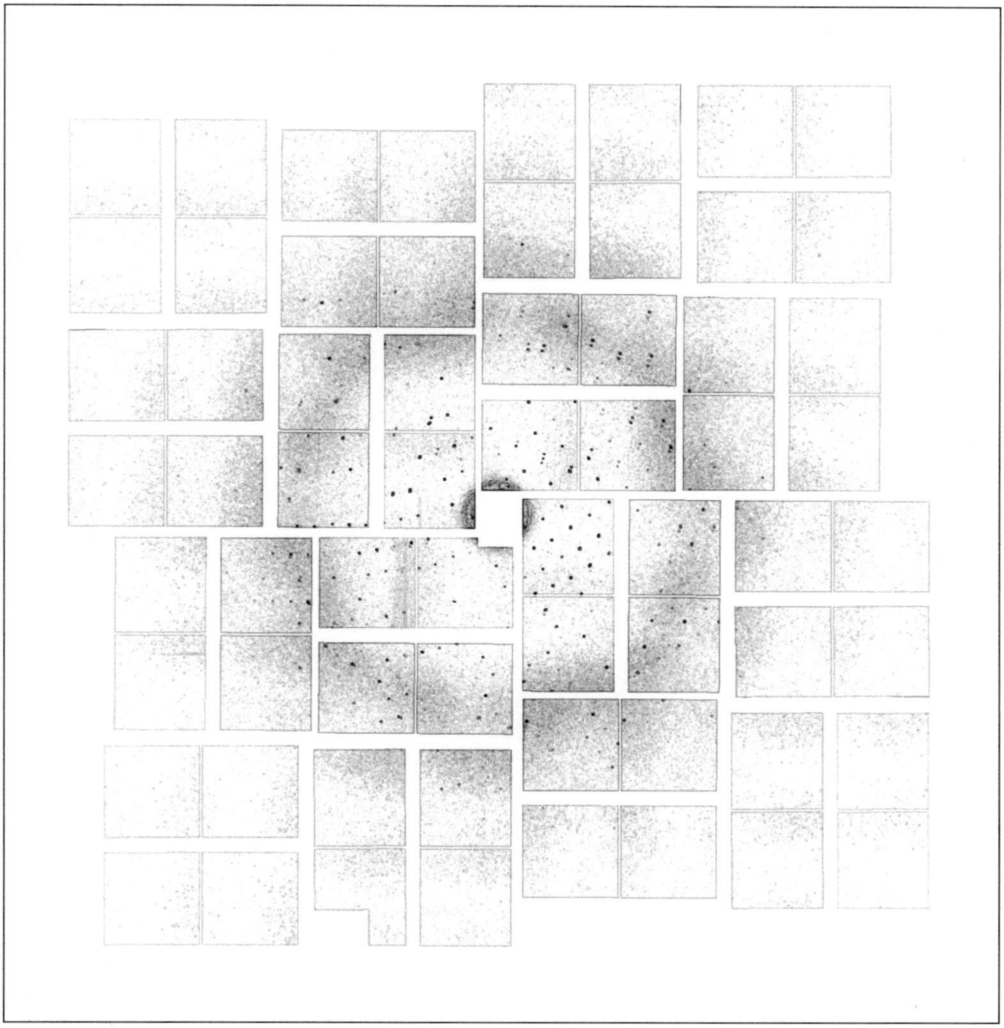

Figure 6.11: Sample image with noise reduction, inverted intensity, and edge detection applied. Now the shape of the Bragg spots is more similar and the contrast has been increased further.

```
procedure CLUSTERDETECTION(image)
Input: The image matrix in which clusters should be detected.
Output: A matrix clusterMatrix containing all clusters found.
  clusterMatrix ← dim(image)
  maxDistance ← 10
  brightnessThreshold ← 100
  for x = 0 + maxDistance; x < size(image) - maxDistance; x++ do
    for y = 0 + maxDistance; y < size(image) - maxDistance; y++ do
      for i = x - maxDistance; i + maxDistance; i++ do
        for j = y - maxDistance; j + maxDistance; j++ do
          if image[x][y] >brightnessThreshold
             && calculateDistance(image[x][y],clusterMatrix[x][y]) <maxDistance then
            clusterMatrix[x][y] ← clusterMatrix[x][y] + 1
          end if
        end for
      end for
    end for
  end for
  return clusterMatrix
end procedure
```

Figure 6.12: Cluster detection in pseudo–code

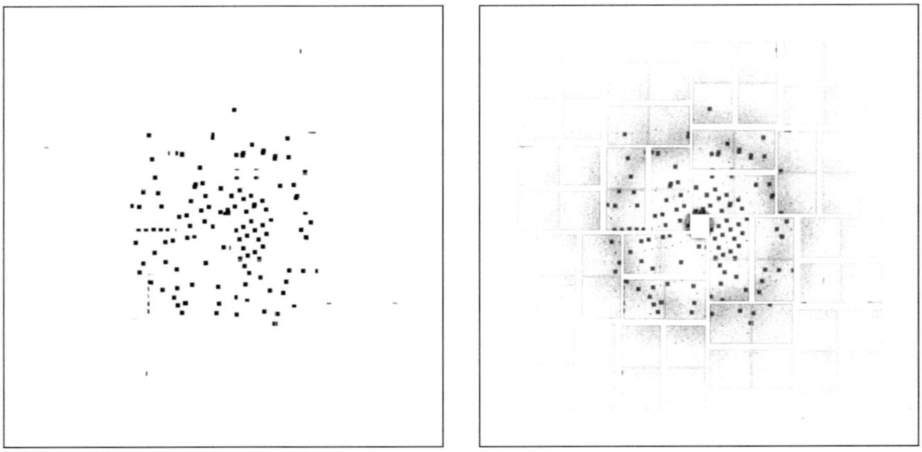

(a) Detected clusters only **(b)** Merged with the original image.

Figure 6.13: Bragg spots found (6.13a) and merged with the original image (6.13b). Most of the spots have been found. Some weaker ones were not identified due to a pessimistically set threshold.

6.2.3 Signal Identification

Once the noise within the image has been removed and the Bragg spots are isolated, it is possible to detect them using a simple, iterative algorithm. We define a Bragg spot as a cluster of bright pixels whose total intensity is above threshold of 100. The pseudocode of the algorithm is shown in Figure 6.12.

The algorithm first creates a new matrix with the same dimensions as the input image. All components of the new matrix are set to zero. Then, a maximum distance and brightness threshold is defined. Next, two nested for-loops iterate over the whole image in x- and y- direction. Within these loops, two additional for-loops examine the pixels up to the maximum distance around the current pixel. If the pixel analyzed is above the brightness threshold, the surrounding pixels are updated. Given the distance between the surrounding pixel and the current pixel is lower than the defined maximum, the corresponding coordinates x, y of the new matrix is increased by 1. After iterating over each pixel of the source image, the new matrix contains all clusters found in the input image.

Since the main factor for this algorithm is the image size, its complexity is $\mathcal{O}(n^2)$. The spots detected by this algorithm can be seen in Figure 6.13a. Figure 6.13b shows an overlay of the spots found with the original, normalized image (Figure 6.9).

6.3 Optimizations

6.3.1 Binary Edge Detection

After edge detection has been applied to an image, all subsequent steps in the analysis depend on a constant intensity threshold. This reduces all comparisons to a binary decision. Therefore, this threshold can directly be applied after edge detection has been applied to a pixel. This can be done by introducing the step function

$$\theta_{\text{threshold}}(I_i) = \begin{cases} 0, I_i < \text{threshold} \\ 1, I_i \geq \text{threshold} \end{cases} \tag{6.16}$$

where I_i denotes the intensity of the i-th pixel.

In subsequent steps, all pixels with the value 1 will always contain data and no further check against a threshold is necessary. In addition, since the valid spots are singled out already, there is no need for a separate cluster detection step.

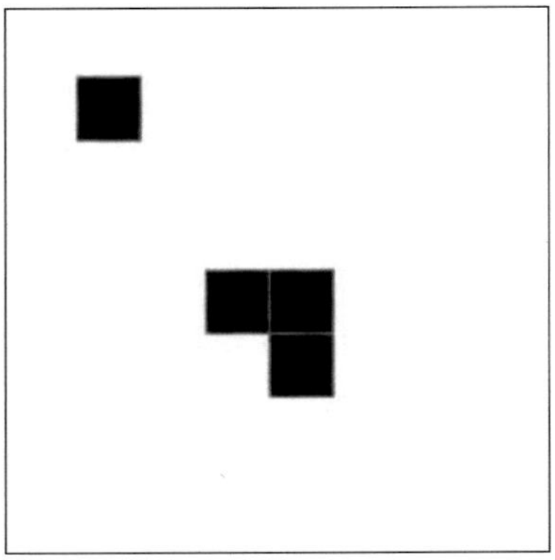

Figure 6.14: Example for single pixel noise. The top left black pixel can be removed, whereas the three connected pixels represent a valid signal.

6.3.2 Single Pixel Removal

After all optimization steps have been applied, an isolated pixel can never represent a valid signal. Therefore it is useful to remove single pixels with an intensity above the set threshold. To achieve that, for each pixel above the threshold, all adjacent pixels are checked for intensities above the threshold. If none of the eight checks returns true, the pixel is removed. Figure 6.14 shows an example for single pixel noise. Here, the top left black pixel can be removed, whereas the three connected black pixels should be kept.

6.3.3 Handling of Panel Boundaries

Pixels at panel boundaries need a special handling. Usually they are not taken into account in the analysis. When edge detection is applied to these areas, an edge is recorded due to the high change in in intensity. This may result in a false positive signals.

There are different options for dealing with this effect. A general solution would consist of introducing a matrix of the same size as the image. All of its components that should be ignored are set to 1. This matrix can then be used after edge detection has been applied. Each pixel in the image for which the corresponding matrix component equals 1 is set to 0 in the processed image. That effectively removes all detected edges at boundaries known not to record useful data.

6.3.4 Alternative Edge Detection Operators

Within the scope of our research the well known Sobel operator is used for edge detection. One advantage of the Sobel operator is the simple computation necessary compared to more complex techniques like the Deriche edge detector [40] which requires multiple steps to apply edge detection for an image.

There are many alternative operators which apply a different set of image kernels to an image. Some examples are Prewitt operator [71], Roberts cross operator [57] and the Scharr operator [81]. It should be explored, if these alternative operators could be utilized to improve edge detection further in these kind of images.

6.4 Results

In order to verify our clusterfinder algorithm, 25 random indexable (useful) and non-indexable (not useful) images were selected for each sample (see Section 4.1). For each image, Bragg spots are detected and their coordinates are recorded. The coordinates are then compared against the list of Bragg spots Cheetah found with a distance tolerance of 10 pixels. The tolerance of 10 pixels is chosen to allow for small differences in the calculation of the coordinate of a Bragg spot. This is because the calculation of the center of a spot directly depends on all pixels belonging to one spot. In a second run, we also applied the background subtraction technique, introduced in Section 5.4.1.

The results are shown in Table 6.1 and Table 6.2 without background subtraction and in Table 6.3 and Table 6.4 with background subtraction applied.

In general, it can be seen that with background subtraction applied the amount of spots in common is increased. This is due to a better signal-to-noise ratio of Bragg spots and their surroundings. In case of the CatB and 5HT-2B sample, our algorithm is able to locate more Bragg spots than Cheetah. For GV however, Cheetah identified more Bragg spots. The difference between GV and the other two samples might be explained by the comparably low signal-to-noise ratio in case of the GV sample. It is well-known in signal processing that signals are hard to identify if the signal-to-noise ratio is small. Consequently, it cannot be expected that an 'optimal' and 'stable' threshold can be specified which allows to identify as many Bragg spots as possible while minimizing false positives.

	Total spots: Cheetah	Total spots: clusterfinder	Spots in common[1]	Spots: Cheetah only[1]	Spots: clusterfinder only[1]
CatB	937	1532	89 %	11 %	48 %
5HT-2B	1400	2169	68 %	30 %	59 %
GV	1180	1603	42 %	57 %	60 %

Table 6.1: Spots identified in indexable images. [1]Compared to Cheetah.

	Total spots: Cheetah	Total spots: clusterfinder	Spots in common[1]	Spots: Cheetah only[1]	Spots: clusterfinder only[1]
CatB	0	66	0 %	0 %	100 %
5HT-2B	827	1768	68 %	30 %	59 %
GV	2015	1287	45 %	54 %	53 %

Table 6.2: Spots found in non-indexable images. [1]Compared to Cheetah.

	Total spots: Cheetah	Total spots: clusterfinder	Spots in common[1]	Spots: Cheetah only[1]	Spots: clusterfinder only[1]
CatB	937	1427	90 %	10 %	42 %
5HT-2B	1400	2169	72 %	27 %	58 %
GV	1180	1093	43 %	56 %	19 %

Table 6.3: Spots found in indexable images using background subtraction. [1]Compared to Cheetah.

	Total spots: Cheetah	Total spots: clusterfinder	Spots in common[1]	Spots: Cheetah only[1]	Spots: clusterfinder only[1]
CatB	0	22	0 %	0 %	100 %
5HT-2B	827	1768	59 %	39 %	68 %
GV	2015	1204	33 %	67 %	14 %

Table 6.4: Spots found in non-indexable images using background subtraction. [1]Compared to Cheetah.

6.5 Summary

In this chapter we tried to answer the question, why sophisticated state of the art algorithms for noise removal can not be used to remove noise in images from nanocrystallography.

We also introduced a multi-step algorithm, able to identify Bragg spots within an image. To achieve this, firstly, single pixel noise within the image is dampened by distributing the intensity of single bright pixels to adjacent pixels. Secondly, fluctuations in intensity over the area of a few pixels are enhanced using edge detection. Thirdly, all connected pixels above a given threshold are marked by a cluster finding algorithm.

Using our algorithm, we were able to identify up to $90\,\%$ of the signals Cheetah found. Our clusterfinder algorithm also identified additional signals.

In addition we introduced several optimizations. Binary edge detection combines the previously mentioned edge detection with thresholding, which can be build upon in later steps, eliminating the need for a dedicated cluster detection step. Single pixel noise removal as well as empty grid removal are introduced as well, helping to reduce false positives. This is done by removing single bright pixels after edge detection has been applied. In addition we proposed a handling of panel boundaries by defining a matrix with information about pixels that should be ignored since they do not record valid data and may contribute to false positives.

Chapter 7

Performance Aspects

In this chapter, we address the research question **3. Is it possible to solve the previous two questions within real-time constraints?** We present a prototypical implementation of our work discussed in Chapter 5 and 6. The findings have been published in article [22].

Parts of the prototype are implemented in OpenCL utilizing a GPU as an accelerator device. The prototype is discussed in terms of its implementation, recognition rate, and runtime behavior. At the end, we draw conclusions on the implication of our benchmarks in nanocrystallography.

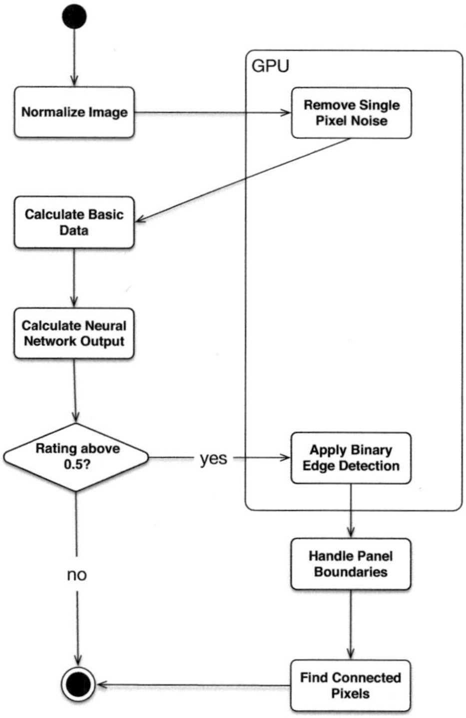

Figure 7.1: State diagram of the proposed prototype. Firstly, the prototype normalizes and optimizes the image. Secondly, our neural network is used to categorize an image. Provided an image is rated above 0.5, then, thirdly, it is considered as containing data. Edge detection is applied and the coordinates of the signals (Bragg spots) are determined.

7.1 Prototypical Implementation

To verify our findings, we implemented a prototype of our proposed algorithms. Previously, the algorithms have been implemented independently in Java to verify their capabilities. To explore the performance and runtime behavior of our algorithms, we implemented our prototype in C. It is composed of a combination of a modified version of the neural network discussed in Section 5.2 as well as our clusterfinder algorithm discussed in Chapter 6. We include several functions to be carried out by an accelerator device such as a GPU. The prototype is implemented in multiple steps as shown in Figure 7.1. In the following paragraphs, each step is described in detail.

Normalize Image Due to technical defects, individual pixels may show a value outside the range the detector is recording (14 bit). Therefore, all pixels outside the range of $0 < I_i < 16{,}364$ are set to 0.

Remove Single Pixel Noise As discussed in Section 6.2.1, most of the noise within images is composed of individual bright pixels. Therefore, all subsequent steps in the analysis chain can profit from removing this noise as soon as possible.

Since the output for each pixel in this step only depends on the source image, it can be efficiently parallelized on an accelerator device.

To exploit the parallel processing capabilities of accelerator devices supported by OpenCL, as many independent work packages as possible should be created (see Section 3.6). Therefore, the processing kernel to be run on the GPU has been assigned two dimensions with 1552 and 1480 items respectively. This represents the dimensions of the image processed and ensures that all execution units are utilized. To save execution time, jump instructions in GPU code should be avoided. Therefore, the commonly used for-loop for calculating the convolution of a pixel with an image kernel by processing all adjacent pixels has been unrolled to improve processing speed. The image kernel used can be seen in Listing 7.1.

The parameters 'image', 'output' and 'width' are passed to the image kernel. Since all of them are shared by all work items they have to be in global memory. Image as well as width are read-only, which means they can be declared as constant, improving the latency for accessing them. In line 6, the variable column stores the current column of the processed image by retrieving the current value for dimension 0. In the next line, the variable pixel is calculated. It represents the actual coordinate of the pixel processed by this work item. It is derived by retrieving the current value for dimension 1 multiplied by the passed width plus the previously calculated column offset.

In line 12 the image kernel is defined. Since all elements are the same, it is defined as the value of one component. It is the same image kernel proposed in Section 6.2.1 for dampening single pixel noise. In the lines 19 to 27, the values for the current pixel as well as all adjacent ones are calculated by multiplying their value with the filter and calculating their sum. Since more complex logic like loops are executed rather slowly on accelerator devices, the loop iterating over the 9 values has been unrolled. The last line finally sets the output of the current pixel to $\frac{1}{9}$ of the calculated sum to avoid an overflow of the short variable.

```
1  __kernel void convolute_unrolled(
2      const __global short *image,
3      __global short *output,
4      const __global short *width) {
5
6      // The pixel offset we are working on is calculated
7      int column = get_global_id(0) + 1;
8      int pixel = (get_global_id(1) + 1) * *width + column;
9
10     // The kernel is set as one value since all elements
11     // are the same
12     float kernel = 1/9.0;
13
14     // Initialize the accumulator with 0
15     float accumulator = 0;
16
17     // Unrolled for-loop summing the 3x3 matrix with the
18     // current pixel at its center
19     accumulator += image[pixel - 1 - *width] * kernel;
20     accumulator += image[pixel - *width] * kernel;
21     accumulator += image[pixel + 1 - *width] * kernel;
22     accumulator += image[pixel - 1 ] * kernel;
23     accumulator += image[pixel] * kernel;
24     accumulator += image[pixel + 1 ] * kernel;
25     accumulator += image[pixel - 1 + *width] * kernel;
26     accumulator += image[pixel + *width] * kernel;
27     accumulator += image[pixel + 1 + *width] * kernel;
28
29     // Set the output pixel to 1/9 of the
30     // computed accumulator to avoid an overflow
31     output[pixel] = (short) (accumulator / 9.0);
32 }
```

Listing 7.1: Convolution carried out on the GPU

Calculate Basic Data In order to classify the data after the convolution is applied, we use a modified version of the neural network proposed in Chapter 5. The quantities I'_{max}, I'_{mean} and $\Delta I'$ are calculated by iterating over the noise reduced image once.

To improve the runtime, only one output neuron is used, which reduces the necessary calculation to

$$O_{rating}(I') = S(I'_{\text{max}} w_1 + I'_{\text{mean}} w_2 + \Delta I' w_3 + w_4) \tag{7.1}$$

where S is the Sigmoid function.

Due to our previous experience with the neural network design introduced in Chapter 5, which has shown that the rating is either close to 0 or close to 1, we choose 0.5 as the threshold between indexable and non-indexable images. We consider all images for which the network outputs a value ≥ 0.5 as indexable and continue with the identification of Bragg spots within the image.

Apply Binary Edge Detection Our optimized binary edge detection is applied to the convoluted image I' (see Section 6.3.1).

Since edge detection also uses convolution, this step has been implemented in OpenCL as well. Similar to the 'Remove Single Pixel Noise' step, an unrolled version of the convolution using the Sobel operator has been implemented as an OpenCL kernel. In addition, an intensity threshold is applied. It is applied once edge detection has been performed. Given an intensity above the threshold, the output is set to 1, otherwise to 0. A listing of the binary edge detection kernel is shown in Listing 7.2.

Various quantities are passed to the execution kernel: the image to be processed, the destination image, the image width as well as an intensity threshold. Except for the output all parameters are set to constant, since they are only read. All variables are also stored in global device memory, since they need to be accessed by all work items. In line 7 the current column is calculated by retrieving the id of the work package for the first dimension. Then, in line 8, the current pixel index is calculated. First, the current id for the second dimension is retrieved. It is then multiplied with the image width in order to retrieve the first index of the current row. The column offset is then added to retrieve the correct index for the current work package. In line 11 and 12 the Sobel operator is defined. It has been introduced in Section 6.2.2.

In the lines 20 to 45, the Sobel operator is applied to the current pixel as well as all adjacent ones. For performance reasons, no loop is used. Line 48 then calculates the result for the current pixel.

Line 52 to 56 apply thresholding to the value just calculated, as discussed in Section 6.3.1.

```
__kernel void binary_sobel(const __global short *image,
__global short *output, const __global short *threshold,
const __global short *width) {
float aX, aY, result;

// Calculate the offset for the current pixel
int column = get_global_id(0) + 1;
int pixel = (get_global_id(1) + 1) * *width + column;

// Define the Sobel operator for x and y direction
short sobel_x[9] = {-1, 0, 1, -2 , 0, 2, -1, 0, 1};
short sobel_y[9] = {-1, -2, -1, 0, 0, 0, 1, 2, 1};

// Initialize the accumulators for x and y direction
aX = 0.0;
aY = 0.0;

// Unrolled for-loop for summing both 3x3 matrices with
// the current pixel at its center
aX += image[pixel - 1 - *width] * sobel_x[0];
aY += image[pixel - 1 - *width] * sobel_y[0];

aX += image[pixel - *width] * sobel_x[1];
aY += image[pixel - *width] * sobel_y[1];

aX += image[pixel + 1 - *width] * sobel_x[2];
aY += image[pixel + 1 - *width] * sobel_y[2];

aX += image[pixel - 1 ] * sobel_x[3];
aY += image[pixel - 1 ] * sobel_y[3];

aX += image[pixel] * sobel_x[4];
aY += image[pixel] * sobel_y[4];

aX += image[pixel + 1] * sobel_x[5];
aY += image[pixel + 1] * sobel_y[5];

aX += image[pixel - 1 + *width] * sobel_x[6];
aY += image[pixel - 1 + *width] * sobel_y[6];

aX += image[pixel + *width] * sobel_x[7];
aY += image[pixel + *width] * sobel_y[7];

aX += image[pixel + 1 + *width] * sobel_x[8];
aY += image[pixel + 1 + *width] * sobel_y[8];

// Result is the combination of both accumulators
result = sqrt(pow(aX, 2) + pow(aY, 2));

// Directly apply thresholding to the data,
```

```
    // set 1 for > threshold, 0 for <= threshold
    if (result > *threshold){
      result = 1;
    } else {
      result = 0;
    }

    output[pixel] = (short) result;
}
```

Listing 7.2: Binary Sobel carried out on the GPU

Handle Panel Boundaries To avoid false positives, a predefined matrix is used to locate and remove pixels at boundaries not recording useful data. This is described in Section 6.3.3. It also saves computational resources, since in the next step, for the removed pixels no adjacent pixels have to be checked. Since, at this stage of the processing, the image matrix only contains binary values, the values of the empty grid matrix have to be binary as well, 1 meaning the pixel should be removed, 0 meaning it should not change. Therefore, for each pixel of the binary image, the corresponding pixel of the empty grid matrix is subtracted. If the value of the source pixel is already 0, no subtraction is necessary.

Find Connected Pixels In the last step, connected pixels are detected using a recursive algorithm. We iterate over each pixel of the image. If a pixel has the value 1, all adjacent pixels are examined recursively until all connected pixels are found. In case of a pixel having the value 1 and no adjacent pixels are set to 1, the pixel is set to 0 and discarded immediately, since a single pixel is very unlikely to be a valid Bragg spot after binary edge detection has been applied. Once the coordinate and intensity of a pixel is recorded, its value is set to 0 to prevent the algorithm from looking at it again. Given the worst case of each pixel within the image is set to 1, the algorithm will perform recursion with a depth of the largest dimension of the matrix. In our case, this would mean a recursion depth of 1580. However, this is only a theoretical case, since a typical Bragg spot is only the size of a few pixels, keeping the recursion rather low. A listing of the relevant code can be seen in Listing 7.3.

Firstly, the currently processed pixel is added to the spot currently processed. Secondly, the pixel just processed is set to 0 in order to avoid handling it again. Thirdly, all adjacent pixels are checked. In a first step, it is examined whether its value is above 0. If so, it is verified that the current coordinate has not already been recorded. Given this is true as well, the function calls itself with the coordinate it just checked in order to add it to the currently processed spot as well.

```
void add_spot_recursive(braggspot *last_spot, int coordinate,
  short *clusters, short *data){

  append_pixel_to_spot(last_spot,
    create_new_braggspot_pixel(coordinate, data[coordinate]));

  // Set processed pixel to 0
  clusters[coordinate] = 0;

  //left
  if (clusters[coordinate - 1] > 0){
    if (!search_for_coordinate_in_braggspot(
      last_spot, coordinate - 1))
        add_spot_recursive(last_spot,
          coordinate - 1, clusters, data);
  }
  // right
  if (clusters[coordinate + 1] > 0){
    if (!search_for_coordinate_in_braggspot(
      last_spot, coordinate + 1))
        add_spot_recursive(last_spot,
          coordinate + 1, clusters, data);
  }
  //top
  if (clusters[coordinate - DIM_X] > 0){
    if (!search_for_coordinate_in_braggspot(
      last_spot, coordinate - DIM_X))
        add_spot_recursive(last_spot,
          coordinate - DIM_X, clusters, data);
  }
  //bottom
  if (clusters[coordinate + DIM_X] > 0){
    if (!search_for_coordinate_in_braggspot(
      last_spot, coordinate + DIM_X))
        add_spot_recursive(last_spot,
          coordinate + DIM_X, clusters, data);
  }
  //top left
  if (clusters[coordinate - DIM_X - 1] > 0){
    if (!search_for_coordinate_in_braggspot(
      last_spot, coordinate - DIM_X - 1))
        add_spot_recursive(last_spot,
         coordinate - DIM_X - 1, clusters, data);
  }
  //top right
  if (clusters[coordinate - DIM_X + 1] > 0){
    if (!search_for_coordinate_in_braggspot(
      last_spot, coordinate - DIM_X + 1))
        add_spot_recursive(last_spot,
         coordinate - DIM_X + 1, clusters, data);
```

```
    }
    //bottom left
    if (clusters[coordinate + DIM_X - 1] > 0){
      if (!search_for_coordinate_in_braggspot(
        last_spot, coordinate + DIM_X - 1))
          add_spot_recursive(last_spot,
            coordinate + DIM_X - 1, clusters, data);
    }
    //bottom right
    if (clusters[coordinate + DIM_X + 1] > 0){
      if (!search_for_coordinate_in_braggspot(
        last_spot, coordinate + DIM_X + 1))
          add_spot_recursive(last_spot,
            coordinate + DIM_X + 1, clusters, data);
    }
}
```

Listing 7.3: Recursive detection of Bragg spots in a binary image.

7.2 Recognition Rates

To verify our prototype, 10 indexable and 10 non-indexable images of each of the three samples (see Section 4.1) have been manually verified and selected. As mentioned in Section 4.2, there is currently no solution capable of identifying all of the signals correctly. To create an objective test set, the images were inspected and categorized manually. In a first step, we categorized the images as 'indexable' or 'non-indexable' based on the calculated output of the neural network. All images for which we received an output value ≥ 0.5 are considered as indexable. To all these images, the clusterfinder algorithm is applied to identify all Bragg Spots, their coordinates and intensities. The spots found are then compared to the ones found using Cheetah (see Section 3.1). The results can be found in Table 7.1 and 7.2. The rather low amount of signals detected by our proposed prototype is due to a conservative intensity threshold, ensuring as few false positives as possible.

It can be seen that besides the signals both algorithms found, each algorithm was also able to detect signals, the other one missed. To explore this behavior, we took two images from each sample and verified the signals found by both applications manually. The results for these two images per sample can be found in Table 7.3. It can be seen that both algorithms find additional valid spots as well as false positives. We found that the false positives in case of the prototype are due to broken pixels within a panel and the high amount of noise around the beam hole. Cheetah, on the other hand, identified the most false positives at panel edges and within the water halo. In addition, we found that the valid spots identified by neither algorithm were almost exclusively very weak. Therefore, their identification depends on the optimizations used before searching for signals, which differs between our prototype and Cheetah.

To summarize, Cheetah and our algorithm differ in their sensitivity in identifying false positive signals. Although a more thorough exploration of the dependence of our algorithm on the parameters is needed, a clear identification should not be expected. Rather, if a signal changes its state from 'signal' to 'non-signal' (or vice versa) when a second run with a somewhat different intensity threshold value is performed, the corresponding signal should be marked as critical and may be investigated more carefully later on in an offline analysis step.

In comparison to the previous recognition rates in Chapter 6 the amount of spots commonly found by Cheetah as well as our prototype is lower. In Table 7.3 it is shown that our prototype generally found less spots than Cheetah. This can be explained by the rather conservatively set threshold. The threshold has been selected to ensure as few false positives as possible. As can be seen in Table 7.3, in case of the 5HT-2B and GV sample, there are still false positives. These are all without exception results of defects of the detector that Cheetah was aware of but our prototype was not.

	Total Spots Found	Spots in Common	Spots Cheetah only	Spots Prototype only
CatB	442	294	118	30
5HT-2B	465	190	207	68
GV	871	265	419	187

Table 7.1: Total spots found in indexable images

	% Spots in common	% Spots Cheetah only	% Spots Prototype only
CatB	67 %	27 %	7 %
5HT-2B	41 %	45 %	15 %
GV	30 %	48 %	21 %

Table 7.2: Percentage of spots found in indexable images

	Total Spots Found	Spots in Common	Spots Cheetah only (valid)	Spots Prototype only (valid)
CatB	36	20 (20)	35 (3)	0
5HT-2B	52	44 (44)	22(15)	11 (6)
GV	95	21 (21)	51(33)	23 (8)

Table 7.3: Bragg spots found in two indexable images for each sample. The numbers in parentheses indicate the number of 'valid' Bragg spots that could be uniquely identified by hand.

7.3 Runtime

Due to the high data rate of experiments to come, it is very important to be able to process images in an efficient way. Therefore, it is necessary to process images in parallel. To explore the capabilities of our prototype, its individual steps have been applied to

- the whole image – 64 panels
- the top and bottom half of the image – 32 panels each
- four individual quarters of the image – 16 panels each
- sixteen individual parts of the image – 4 panels each.

The prototype has been compiled using Apple LLVM version 7.0.2 (clang-700.1.81) on Mac OS 10.11.3. No compiler optimizations have been used. The runtime of the prototype has been measured using the Instruments application 7.2.1 which is part of the Apple Xcode IDE [1]. In Instruments, the 'time profiler' instrument has been used to determine the runtime for each step of the prototype. For each run of the application, one image is processed. In total, 10 runs have been performed for each part of the image. Before executing the benchmark, the system has been rebooted. The default system services are enabled and the only foreground application is Instruments.

The execution time is measured after an image has been loaded into memory. For each step the resulting data, given there are any, are copied into a new matrix. In steps involving the GPU, the data transfer to and from the GPU is included in the measured time as well.

The runtime of all variations can be found in Figure 7.3. The values are taken from Table 7.4 where, in addition, the standard deviation is indicated. For simplicity, it is identified with the largest standard deviation of the partial measurement. This is sufficient to show that the uncertainty of the mean values is sufficiently small, i.e. qualitative conclusions can be drawn from them.

It can be seen that the runtime decreases the fewer panels are analyzed. The decrease tends to saturate for 'Remove Single Pixel Noise' as well as 'Binary Edge Detection'. This is most likely due to the overhead of managing and moving data in memory, since as of now, GPUs do not have access to system memory [80]. We expect an even smaller improvement for the processing of one panel only, since memory management still has to be done.

The speedup of the other partial measurements is increasing nearly linearly with the inverse number of panels. It can also be seen that the average runtime of the 'Rate using Neural Network' step is rather high. The long runtime compared to other steps might be explained by the fact, that in order to calculate the three basic quantities, it is necessary to iterate and sum over each pixel of the image which, in turn, is compatible with the observed $\mathcal{O}(n^2)$ behavior of the runtime. The rather high standard deviation might be explained by the amount of independent steps necessary to calculate the output of the network.

The benchmarks have been run a system whose specification is shown in Table 7.5. The influence of the 'Turbo Boost' technology[32] has not been explored in our research.

In the following we estimate how many images per second can be processed using our algorithm in the realm of photon science on the available hardware. Several options are taken into account. Since the runtime is not decreasing at least linearly and given a constant stream of images it is faster to process a full image (64 panels) in one process. However, we are proposing a solution very close to the detector device. Here, only parts of the image are available. Therefore, the following calculations are carried out using only parts of the full image.

Step	64 Panels		32 Panels		16 Panels		4 Panels	
	AVG	SD	AVG	SD	AVG	SD	AVG	SD
Normalize Image	5.8 ms	0.1 ms	3.0 ms	0.2 ms	1.5 ms	0.1 ms	0.4 ms	0 ms
Remove Single Pixel Noise	15.3 ms	1.2 ms	12.8 ms	0.8 ms	11.8 ms	0.7 ms	11.2 ms	0.4 ms
Rate using NN	15.9 ms	0.8 ms	8.1 ms	0.5 ms	4.0 ms	0.2 ms	1.0 ms	0.1 ms
Binary Edge Detection	10.4 ms	1.4 ms	7.3 ms	0.8 ms	6.3 ms	0.5 ms	5.1 ms	0.4 ms
Find connected Pixels	8.1 ms	1.3 ms	5.6 ms	1.2 ms	2.9 ms	0.2 ms	2.3 ms	0.4 ms
Total	55.5 ms	3.3 ms	37 ms	2.7 ms	26.5 ms	2.8 ms	20 ms	2.1 ms

Table 7.4: Average runtime (AVG) for 10 runs of each step including the standard deviation (SD).

CPU	Intel I7-4650U 1.7 GHz (TurboBoost 3.3 GHz), 2 Cores
GPU	Intel HD Graphics 5000, 1.5 GB VRAM, 40 execution units
RAM	8 GB DDR 1600 MHz

Table 7.5: Technical specs of the system used for benchmarking

Case 1 Given the measured runtimes and taking into account the hardware specs, it would be feasible to run two processes on this system at the same time in parallel where each process is analyzing one half-image each. According to our benchmark from Table 7.4, a half-image is processed in 37 ms. This would result in

$$1 \text{ half-image per CPU} \times 2 \text{ Cores} = \frac{1}{2} \frac{1}{37/1000} 2 \frac{\text{images}}{\text{s}} = 27.03 \frac{\text{images}}{\text{s}} . \tag{7.2}$$

Case 2 State of the art server hardware will be able to process many more images per second. Let us take the system currently used for the analysis of diffraction images in nanocrystallography at the Center for Free Electron Lasers (CFEL) in Hamburg [82] as an example. It has 24 CPUs and 144 physical cores in total. The detailed technical specs are shown in Table 7.6.

Let us further assume that each CPU core is capable of processing some part of an image at least at the speed of our system used for benchmarking and that a GPU with sufficient performance is present. Our benchmark showed that it is possible to process 4 panels in 20 ms, see Figure 7.3. Therefore, $\frac{64}{4} = 16$ processes are needed for processing a whole image. By utilizing all 144 cores in parallel, 450 diffraction images could be processed per second

$$\frac{1}{16} \text{ image per core} \times 144 \text{ cores} = \frac{1}{16} \frac{1}{20/1000} 144 \frac{\text{images}}{\text{s}} = 450 \frac{\text{images}}{\text{s}} \tag{7.3}$$

Case 3 The previous case assumes an efficiency of 100 %, meaning that every diffraction image is indexable.

In real-world experiments, the efficiency is often as low as 5 % or even less and may increase to 50 % [25].

In our prototype for each image the steps

- Normalize Image
- Remove Single Pixel Noise
- Calculate Neural Network Output

are executed, which accumulate to 12.6 ms when 4 panels are analyzed, see Table 7.4. However, the steps

- Apply Binary Edge Detection
- Handle Panel Boundaries
- Find connected Pixels

adding 7.4 ms processing time, are only executed if the image is rated as 'indexable' by the neural network. This means that only the steps up to the point of calculating a rating for the image need to be carried out for each image. In Equation 7.4 it has been shown that 450 images per second could be processed in the case of each image being indexable. If no image contains Bragg spots only the steps up to the rating by the neural network have to be applied, i.e. more images could be processed per second:

$$\frac{1}{16} \text{ image per core} \times 144 \text{ cores} = \frac{1}{16} \frac{1}{12.6/1000} 144 \frac{\text{images}}{\text{s}} = 714.29 \frac{\text{images}}{\text{s}} \tag{7.4}$$

In Figure 7.2 the runtime as a function of the efficiency between 0 % and 100 % is shown by interpolation between 12.6 and 20 ms. The red line shows the average runtime per image in ms. The green line shows the amount of images the system is capable of analyzing per second.

The diagram shows that at a 50 % efficiency, which represents the best real-world case [17], it would be possible to process 545 images per second. In the case of a 5 % efficiency, 696 images could be processed per second.

Case 4 The processing speed could be improved even further by separating the steps necessary for categorization of the data and the signal identification and localization. As detailed in Section 2.2.1, each of the four quadrants of the detector are connected to an FPGA aggregating the data from all panels connected. The article 'A high-performance fully reconfigurable FPGA-based 2D convolution processor' [69] explores high performances implementations of convolution on FPGAs. It shows that the process of image convolution can be carried out efficiently by FPGAs.

If the categorization steps up to the rating by the neural network were to be moved to the FPGAs already in place or to a second FPGA connected serially, it would remove the need for performing these steps later on. Since the FPGAs already add attributes to the data captured, a convoluted version of the image as well as the rating by the neural network could be also added and directly used for subsequent analysis.

This, in turn, would save 12.6 ms execution time of the total analysis duration, as the steps 'Normalize Image', 'Remove Single Pixel Noise' and 'Calculate Neural Network Output' would not be performed on the machine. Only the steps 'Apply Binary Edge Detection', 'Handle Panel Boundaries' and 'Find connected Pixels' would need to be carried out here, which takes 3.3 ms in total to process 4 panels.

$$\frac{1}{16}\text{ image per core} \times 144 \text{ cores} = \frac{1}{16}\,\frac{1}{7.4/1000}\,144\,\frac{\text{images}}{\text{s}} = 1216.2\,\frac{\text{images}}{\text{s}} \tag{7.5}$$

Since $\frac{1216.2}{27,000} = 0.045$, approximately $4.5\,\%$ of the diffraction images could be processed at the European XFEL. This, in turn, means that it would be possible to process all indexable images in real-time provided the efficiency is below $4.5\,\%$.

On the other hand, given a higher efficiency than $4.5\,\%$, i.e. more than $4.5\,\%$ of the diffraction images contain Bragg spots, the system would not be able to handle the data rate. This would mean that more hardware would be needed in order to balance the workload between multiple systems.

Buffering might also come to mind to compensate for temporarily higher efficiencies. The size of one image taken by the CSPad detector is

$$2296960 \text{ Pixel} \times 14\,\frac{\text{Bit}}{\text{Pixel}} = 32157440 \text{ Bit} \tag{7.6}$$

which means 3.84 MByte per image.

By multiplying this with the image repetition rate of the European XFEL, we get

$$27,000\,\frac{\text{Image}}{\text{Second}} \times 3.84\,\frac{\text{MByte}}{\text{Image}} = 103{,}680\,\frac{\text{MByte}}{\text{Second}} \tag{7.7}$$

or $103.7\,\text{GB}$ per second of data. Given this amount of data, it is not possible to buffer the data for more than a few seconds using state of the art hardware. This leaves artificially throttling the image repetition rate or discarding some images as the only option.

CPU	24x Intel(R) Xeon(R) CPU X7542 2.67 GHz 6 Cores
RAM	768 GB

Table 7.6: Technical specs of the system used for analyzing diffraction images at CFEL Hamburg.

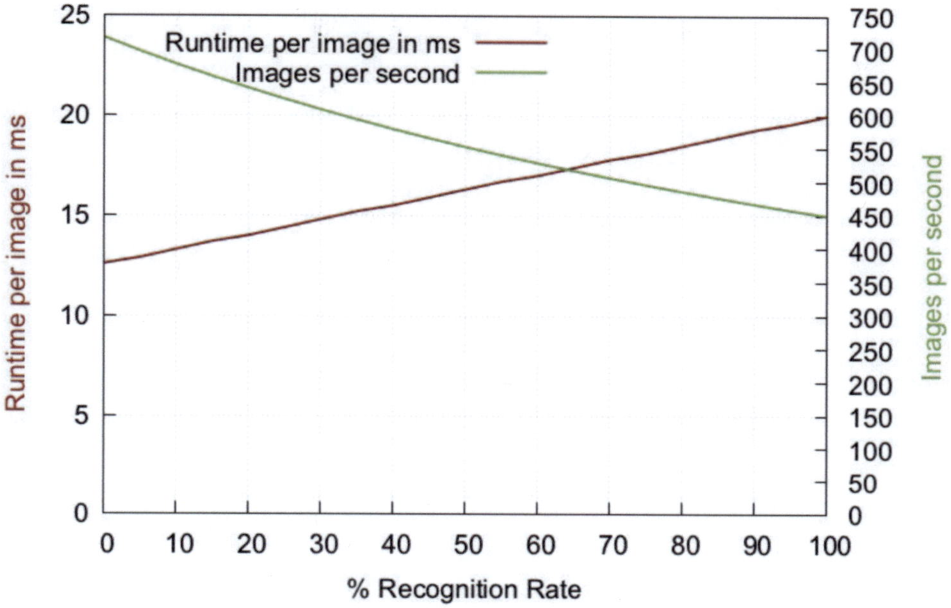

Figure 7.2: Runtime behavior depending on the efficiency of the experiment. The values for 0 % and 100 % have been measured and the steps in between are interpolated. Given a 0 % efficiency, the processing of each image takes 13 ms on average and 690 images can be processed per second. For a 100 % efficiency, the average processing duration is 20 ms per image, leading to 450 images per second.

7.4 Summary

In this Chapter we introduced a prototypical implementation of our proposed algorithms from Chapter 5 and 6. The prototype has been implemented in C. The steps involving convolution of the image are implemented in OpenCL and are carried out on a GPU. Section 7.1 describes the individual steps of the prototype. Next we verified the recognition rate of the prototype by comparing the identified Bragg spots with the ones Cheetah (see Section 3.1) found. We also took two images from each of the three samples and verified each spot found by the prototype as well as Cheetah manually. We found that our prototype as well as Cheetah easily identify Bragg spots with a sufficiently higher photon count compared to the direct surroundings of the spot. The spots exclusively identified by neither our prototype nor Cheetah were only very weak ones, not clearly set apart from their surrounding. Lastly we explored the real-time capabilities of our prototype by measuring the execution time for each main step for the whole image and parts of it. Based on measured execution time of our system benchmarked we then extrapolated the results and laid out a possible analysis setup able to process up to 1216.2 images per second.

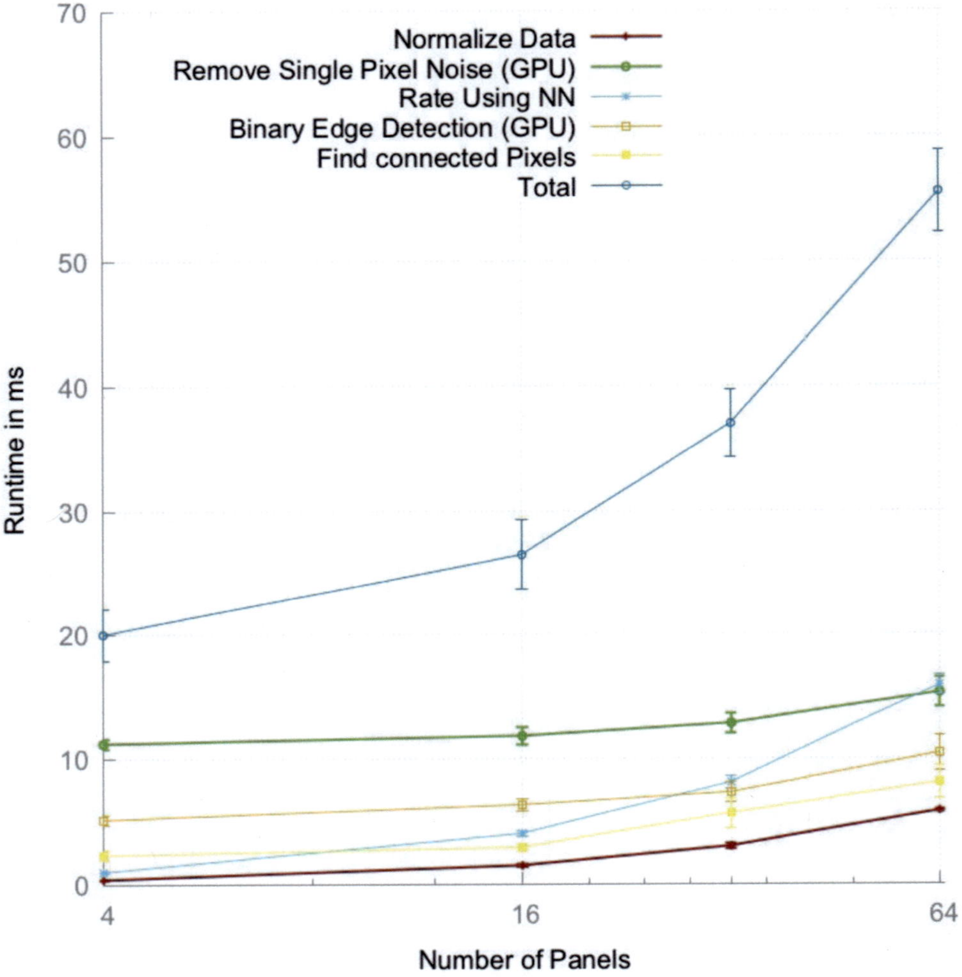

Figure 7.3: Runtime behavior of the prototype for different panel counts. The error bars represent the standard deviation.

Chapter 8
Conclusion

This Chapter concludes this thesis by revisiting the research questions introduced in Chapter 1. The results regarding each question are discussed and the impact of our proposed solutions is evaluated. The main research question this thesis has tried to answer was whether it is possible to design an algorithm capable of rejecting all those images which are useless for further research within the real-time-constraints of current as well as next generation experiments. In the following sections, we answer this question by discussing each research question derived from the main question. This discussion will be followed by an outlook of possible further research.

8.1 How is it possible to determine if there is data within an image at all?

In Chapter 5 we introduced an approach for categorizing images in 'indexable' and 'non-indexable'. The neural network is able to recognize up to 93 % of the images the Cheetah software (see Section 3.1) identified as useful for further research. In order to perform the categorization, only three basic quantities have to be extracted from the image.

We also introduced background subtraction to reduce noise within the images before the analysis. In addition, we introduced the 'transverse intensity' as a new quantity in nanocrystallography images. The transverse intensity can be used to calculate a factor for the compensation of the loss of intensity towards the outer areas of the detector and therefore increases the signals in that area.

In conclusion, it is possible to determine whether an image contains data in a fast and efficient way with the limitation of a sufficiently high signal-to-noise ratio.

8.2 Is the data within an image useful for further analysis?

Chapter 6 discusses an algorithm developed to detect Bragg spots within an image. It consists of multiple steps to remove as much noise as possible before valid signals are identified. Firstly, single bright pixels are dampened using convolution. Secondly, signals are enhanced using edge detection by applying the Sobel operator to the image. And Thirdly, clusters are detected throughout the image using an iterative approach marking all spots above a pre-defined threshold.

The results have been compared against Cheetah as well. We found that we were able to identify up to 90 % of the signals, Cheetah found as well as additional ones.

To sum up this section, we were able to create an algorithm able to identify valid data within the image along with their location. This can be used in steps further down the analysis chain to verify the eligibility of the image for the indexing process by tools like CrystFEL (see Section 3.2).

8.3 Is it possible to achieve the prior two questions with regards to the real-time demands?

In Chapter 7 we presented results obtained from a prototypical implementation of the proposed algorithms in Chapter 5 and 6. In order to increase the performance, parts of the prototype are executed on a GPU rather than the CPU. All major steps of the prototype are explained and the recognition rate is determined. In addition benchmark tests are performed. Based on the numbers gained we estimate the potential of the algorithms to be used in realtime or near-realtime. We found that it might be possible to analyze in realtime the output of the European XFEL experiment, which is designed to produce 27,000 images per second, provided no more than 4.5 % of the images contain useful data which is a typical efficiency value in current experiments. The statement relies on a transfer of the image categorization process to dedicated accelerator hardware. Furthermore, the statement is obtained by considering the computational power of a system currently in use to process diffraction images at CFEL Hamburg. If more potent hardware were to be available the throughput could even be increased further.

In summary it seems to be possible to achieve the categorization of images and identification of signals within each image with regards to real-time demands posed by the European XFEL experiment with certain limitations.

8.4 Can existing algorithms be used to facilitate the image optimization?

There are plenty of algorithms able to remove noise from images and optimize their appearance. However, Bragg spots tend to have a very small extension in diffraction images and are very similar to generic noise. In Chapter 6 we were able to successfully use the technique of convolution to remove single pixel noise from images. We have also shown in Section 6.1, that modern sophisticated algorithms like Block-Matching 3D (see Section 3.8) are not capable of separating noise from signals in diffraction images. This is due to their approach of converting the image into the Fourier space and applying a hard threshold cut by which also small signals are removed from the image. And this, in turn, would also remove valid signals (Bragg spots) which have to be kept.

The outcome of this research question is a perfect illustration of a successful cooperation between informatics and physics. An essential prerequisite was to understand first of all the essence of the physical process (here, that the data in photon science are essentially taken in the Fourier space) and then, in a second step, to choose the right algorithms out of the pool of informatics.

8.5 Design Proposal

A feasible way to process the huge amount of data created by the European XFEL is suggested in Figure 8.2. Given the CSPad detector will be used at the European XFEL, we propose to attach an accelerator system to each panel of the detector. They acquire the data from the panels and, firstly, normalize the data (as described in Section 4.3) using a simple analog cut function before the input is digitized. Secondly, a rating of the data is done by using the neural network proposed in this thesis. Depending on the rating, the data for each panel are either passed to the next level or discarded directly. On a per-quadrant level, another set of accelerators could then remove noise by applying a convolution and rate the data again in order to refine the selection process. As we have shown, removing single pixel noise greatly improves the signal-to-noise ratio of diffraction images. Therefore, a much higher recognition rate can be expected from the neural network, leading to only a small fraction of false positives. All quadrants containing actual data pass their output to the data acquisition module which combines all quadrants to one connected image. This image is then be passed on to a system carrying out the signal identification as well as storing the data.

As an accelerator system, either GPUs or FPGAs might be used. The article ‘Image Convolution Processing: a GPU versus FPGA Comparison’ [79] shows a benchmark of convolutions carried out on CPUs, GPUs and FPGAs. The authors show that using a desktop-class GPU with CUDA support, it is possible to perform a convolution to a 1600 x 1200 pixel image in $0.684\,\mathrm{ms}$.

The runtime behavior of convolution should scale almost linearly with the number of pixels. We assume that this behavior will continue for smaller image dimensions. Each panel of the CSPad detector (see Section 2.2.1) consists of 194 x 185 pixels.

Consequently, the time for performing a convolution is approximately given by

$$\frac{194 \times 185\,\mathrm{pixel}}{1600 \times 1200\,\mathrm{pixel}}\,0.684\,\mathrm{ms} = 0.0128\,\mathrm{ms} \tag{8.1}$$

Phrased differently, $\frac{1}{0.0128} = 77810$ images per second can be convoluted which is much more than experiments like the European XFEL are able to produce (see Chapter 4). Furthermore, it should be noted that the rates obtained in our benchmarks may be improved further by using modern FPGAs in addition to GPUs as accelerator devices.

8.6 Outlook

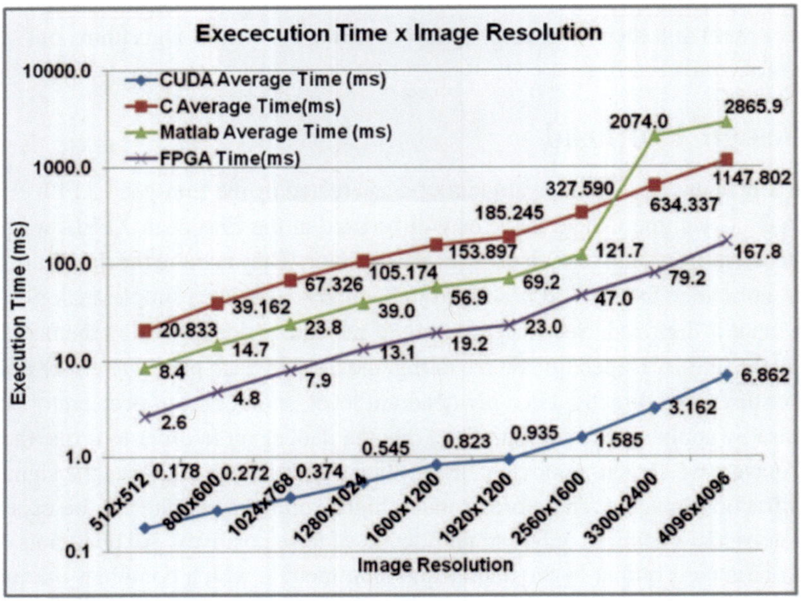

Figure 8.1: Comparison of the execution time of convolution for different image sized and accelerator devices [79]. The red and green lines show implementations on the CPU using C and Matlab. The purple line shows the execution time for an FPGA accelerator and the blue lines shows the runtime for an implementation using CUDA on an NVIDIA desktop class GPU.

The rate at which the European XFEL experiment is able to take data is very challenging. It puts high demands on the performance of possible solutions separating useful data from useless. The algorithms and techniques proposed in this theses can be a foundation for a framework or workflow dealing with this challenge in an efficient way. In addition, newer generations of hardware will be able to process and access data in memory even faster, thus, further increasing the processing capabilities of our solutions. The prototype developed for the context of this thesis could also be improved. As of now, memory is allocated to store the outcome of each step in order to extract partial results. By using only one dataset and only passing pointers, memory copy and allocation could be mostly avoided.

It should also be investigated, whether alternative approaches to our veto engine work in a more efficient way. Instead of using neural networks for classifying diffraction images, different classification algorithms may be explored, for example decision trees or support vector machines.

Our algorithm for detecting Bragg spots within an image might be optimized as well. It could be possible, that an alternative operator for edge detection might help to increase the signal-to-noise ratio of signals even further. In addition, the removal of single pixel noise and edge detection could be carried out in one step, saving execution time, since convolution is associative.

Furthermore, the runtime complexity of the signal identification might be reducible from $\mathcal{O}(n^2)$ to, perhaps, $\mathcal{O}(n \log n)$ by employing a divide and conquer approach. An image could be trimmed until a Bragg spot is reached at each side. The image would then be split in half and the process repeated. This would be done until each sub-image only contains a single Bragg spot, since it does not make sense to go smaller than the magnitude of Bragg spots. Once this step is done, the coordinates of each spot can be returned recursively. However, it would have to be researched how exactly an image would have to be split into sub-images, which might be hard for spots of different shape.

It might also be possible that a different order and combination of the optimizations proposed in this thesis might yield better results. In order to verify this, a test suite could be compiled, able to apply optimizations automatically in any given order. This could be used to automatically find the ideal order and types of optimization leading to the best possible results.

All algorithms presented in this thesis rely heavily on the ability to determine and remove noise. We have only scratched the tip of the iceberg. A more thorough understanding of noise needs a deeper understanding of the physics of the samples under investigation. In order to move forward in this field, the inspiring communication between physicists and computer scientists should be continued.

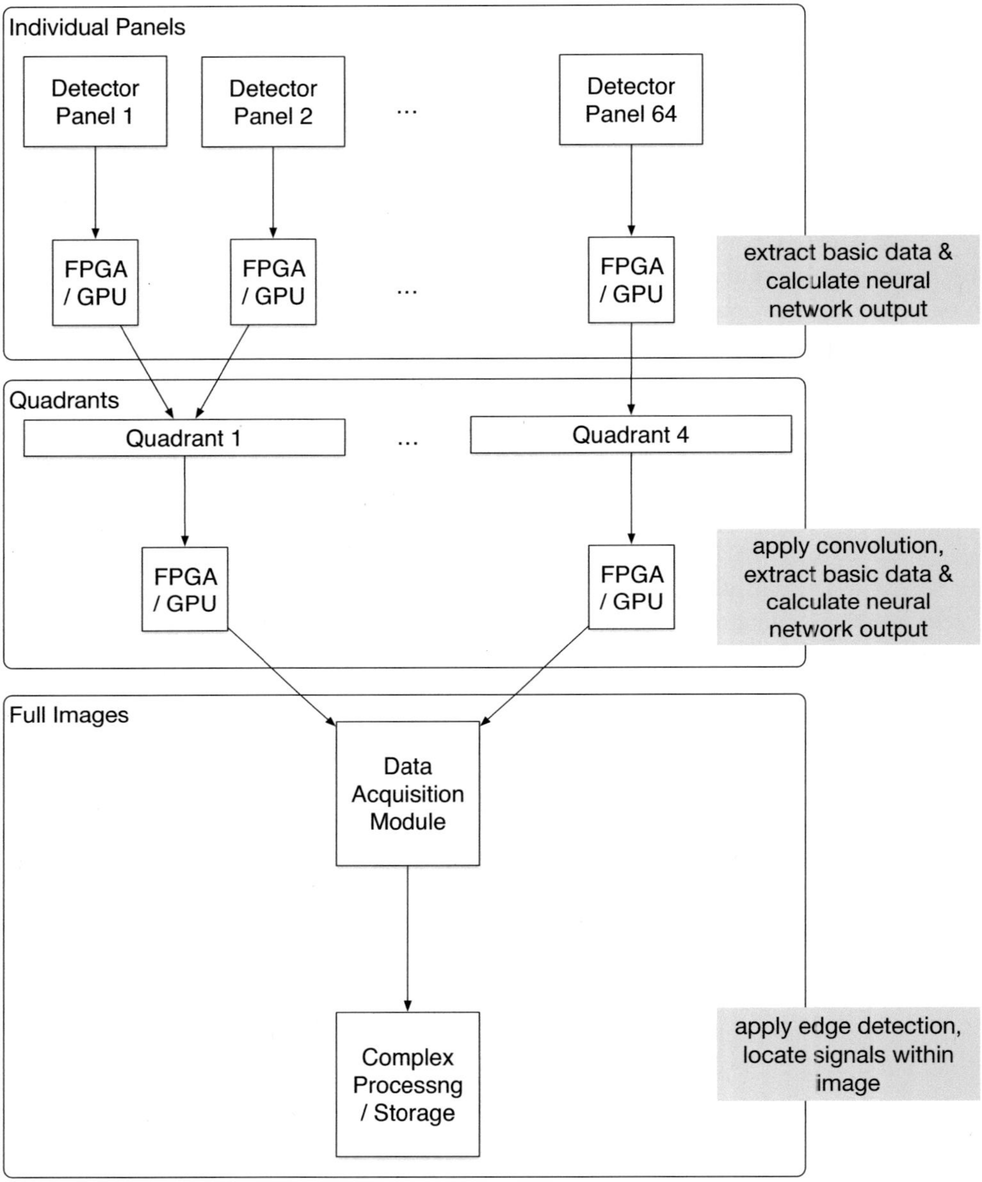

Figure 8.2: Proposed setup for real-time signal identification at the European XFEL.

Bibliography

[1] Apple Instruments. https://developer.apple.com/library/mac/documentation/DeveloperTools/Conceptual/InstrumentsUserGuide/index.html#//apple_ref/doc/uid/TP40004652. [Online; accessed 2015-02-16].

[2] Diffraction of waves by an obstacle. https://upload.wikimedia.org/wikipedia/commons/0/0a/5wavelength%3Dslitwidthsprectrum.gif. [Online; accessed 2015-01-31].

[3] Image of the CSPad detector. https://portal.slac.stanford.edu/sites/lcls_public/instruments/cxi/pictures/Forms/DispForm.aspx?ID=4. [Online; accessed 2015-01-28].

[4] Imagine a crystal's inner life. https://aicr2014.univ-rennes1.fr/pdf/Nature_milestone_august2014.pdf. [Online; accessed 2015-02-13].

[5] Khronos Group. https://www.khronos.org. [Online; accessed 2015-01-25].

[6] Neural network using a single layer. https://commons.wikimedia.org/wiki/File:SingleLayerNeuralNetwork_english.png. [Online; accessed 2015-02-04].

[7] Neural Network with a hidden layer. https://commons.wikimedia.org/wiki/File:MultiLayerNeuralNetwork_english.png. [Online; accessed 2015-02-04].

[8] OpenCL program execution flow. https://upload.wikimedia.org/wikipedia/commons/f/fd/OpenCL_program_execution_flow.jpg. [Online; accessed 2015-02-01].

[9] Optimization of Data Life Cycles. https://indico.cern.ch/event/214784/session/5/contribution/28/attachments/340903/475737/Optimization_of_data_life_cycles_final.pdf. [Online; accessed 2015-02-12].

[10] Recurrent neural network. https://commons.wikimedia.org/wiki/File:RecurrentLayerNeuralNetwork_english.png. [Online; accessed 2015-02-04].

[11] Three-dimensional view of the collected Fourier-coefficients. https://desy.cfel.de/cid/research/serial_femtosecond_crystallography/. [Online; accessed 2015-12-13].

[12] Iftikhar Ahmad, MA Ansari, and Sajjad Mohsin. Performance comparison between backpropagation algorithms applied to intrusion detection in computer network systems. In

Proceedings of the 7th WSEAS International Conference on Applied Computer and Applied Computational Science, pages 47–52. World Scientific and Engineering Academy and Society (WSEAS), 2008.

[13] Imtiaz Ahmad and A Shoba Das. Hardware implementation analysis of sha-256 and sha-512 algorithms on fpgas. *Computers & Electrical Engineering*, 31(6):345–360, 2005.

[14] Jens Als-Nielsen and Des McMorrow. *Elements of modern X-ray physics*. John Wiley & Sons, 2011.

[15] M Altarelli. The european x-ray free-electron laser facility in hamburg. *Nuclear Instruments and Methods in Physics Research Section B: Beam Interactions with Materials and Atoms*, 269(24):2845–2849, 2011.

[16] Thomas Back, David B Fogel, and Zbigniew Michalewicz. *Handbook of evolutionary computation*. IOP Publishing Ltd., 1997.

[17] Anton Barty, Richard A Kirian, Filipe RNC Maia, Max Hantke, Chun Hong Yoon, Thomas A White, and Henry Chapman. Cheetah: software for high-throughput reduction and analysis of serial femtosecond x-ray diffraction data. *Journal of Applied Crystallography*, 47(3):1118–1131, 2014.

[18] Daniel Becker and Achim Streit. A Neural Network Based Pre-Selection of Big Data in Photon Science. *BDCloud*, pages 71–76, 2014.

[19] Daniel Becker and Achim Streit. Localization of signal peaks in photon science imaging. *UKSim*, pages 296–301, 2015.

[20] Daniel Becker and Achim Streit. Real-time signal identification in big data streams Bragg-Spot localization in photon science. *HPCS*, pages 611–616, 2015.

[21] Daniel Becker and Achim Streit. Real-time Signal identification in Photon Science Imaging. *IJSSST*, 2016.

[22] Daniel Becker and Achim Streit. Realtime-processing of nanocrystallography images. *UKSim*, pages 190–195, 2016.

[23] A Berntson, V Stojanoff, and H Takai. Application of a neural network in high-throughput protein crystallography. *Journal of synchrotron radiation*, 10(6):445–449, 2003.

[24] John Billingsley, Kuniaki Kawabata, Mutsunori Takahashi, Kanako Saitoh, Mitsuaki Sugahara, Hajime Asama, Taketoshi Mishima, and Masashi Miyano. Crystalline object evaluation by image processing. *Sensor Review*, 28(2):143–149, 2008.

[25] Sébastien Boutet, Lukas Lomb, Garth J Williams, Thomas RM Barends, Andrew Aquila, R Bruce Doak, Uwe Weierstall, Daniel P DePonte, Jan Steinbrener, Robert L Shoeman,

et al. High-resolution protein structure determination by serial femtosecond crystallography. *Science*, 337(6092):362–364, 2012.

[26] William Henry Bragg and William Lawrence Bragg. The reflection of x-rays by crystals. *Proceedings of the Royal Society of London. Series A, Containing Papers of a Mathematical and Physical Character*, 88(605):428–438, 1913.

[27] H Braun and M Riedmiller. Rprop: a fast adaptive learning algorithm. In *Proceedings of the International Symposium on Computer and Information Science VII*, 1992.

[28] G A Carini, S Boutet, M Chollet, A Dragone, G Haller, P A Hart, S C Herrmann, C J Kenney, J Koglin, H T Lemke, M Messerschmidt, S Nelson, J Pines, A Robert, S Song, J B Thayer, G J Williams, and D Zhu. Measurements at synchrotrons and FELs: Some differences observed with the CSPAD. In *Nuclear Science Symposium and Medical Imaging Conference (NSS/MIC), 2013 IEEE*, pages 1–5. IEEE, 2013.

[29] G A Carini, S Boutet, M Chollet, A Dragone, G Haller, P A Hart, S C Herrmann, C J Kenney, J Koglin, M Messerschmidt, S Nelson, J Pines, A Robert, S Song, J B Thayer, G J Williams, and D Zhu. Experience with the CSPAD during dedicated detector runs at LCLS. *Journal of Physics: Conference Series*, 493(1):012011–4, March 2014.

[30] Henry N Chapman, Anton Barty, Michael J Bogan, Sébastien Boutet, Matthias Frank, Stefan P Hau-Riege, Stefano Marchesini, Bruce W Woods, Saša Bajt, W Henry Benner, et al. Femtosecond diffractive imaging with a soft-x-ray free-electron laser. *Nature Physics*, 2(12):839–843, 2006.

[31] Henry N Chapman, Petra Fromme, Anton Barty, Thomas A White, Richard A Kirian, Andrew Aquila, Mark S Hunter, Joachim Schulz, Daniel P DePonte, Uwe Weierstall, R Bruce Doak, Filipe R N C Maia, Andrew V Martin, Ilme Schlichting, Lukas Lomb, Nicola Coppola, Robert L Shoeman, Sascha W Epp, Robert Hartmann, Daniel Rolles, Artem Rudenko, Lutz Foucar, Nils Kimmel, Georg Weidenspointner, Peter Holl, Mengning Liang, Miriam Barthelmess, Carl Caleman, Sébastien Boutet, Michael J Bogan, Jacek Krzywinski, Christoph Bostedt, Saša Bajt, Lars Gumprecht, Benedikt Rudek, Benjamin Erk, Carlo Schmidt, André Hömke, Christian Reich, Daniel Pietschner, Lothar Strüder, Günter Hauser, Hubert Gorke, Joachim Ullrich, Sven Herrmann, Gerhard Schaller, Florian Schopper, Heike Soltau, Kai-Uwe Kühnel, Marc Messerschmidt, John D Bozek, Stefan P Hau-Riege, Matthias Frank, Christina Y Hampton, Raymond G Sierra, Dmitri Starodub, Garth J Williams, Janos Hajdu, Nicusor Timneanu, M Marvin Seibert, Jakob Andreasson, Andrea Rocker, Olof Jönsson, Martin Svenda, Stephan Stern, Karol Nass, Robert Andritschke, Claus-Dieter Schröter, Faton Krasniqi, Mario Bott, Kevin E Schmidt, Xiaoyu Wang, Ingo Grotjohann, James M Holton, Thomas R M Barends, Richard Neutze, Stefano Marchesini, Raimund Fromme, Sebastian Schorb, Daniela Rupp, Marcus Adolph, Tais Gorkhover, Inger Andersson, Helmut Hirsemann, Guillaume Potdevin, Heinz Graaf-

sma, Björn Nilsson, and John C H Spence. Femtosecond X-ray protein nanocrystallography. *Nature*, 470(7332):73–77, February 2011.

[32] James Charles, Preet Jassi, N S Ananth, Abbas Sadat, and Alexandra Fedorova. *Evaluation of the Intel Core i7 Turbo Boost feature*. IEEE, 2009.

[33] Elaine Chiu, Fasséli Coulibaly, and Peter Metcalf. Insect virus polyhedra, infectious protein crystals that contain virus particles. *Current opinion in structural biology*, 22(2):234–240, 2012.

[34] Paweł Chodowiec and Kris Gaj. Very compact fpga implementation of the aes algorithm. In *Cryptographic Hardware and Embedded Systems-CHES 2003*, pages 319–333. Springer, 2003.

[35] Dan Claudiu Ciresan, Ueli Meier, Luca Maria Gambardella, and Jürgen Schmidhuber. Deep big simple neural nets excel on handwritten digit recognition, 2010. *Cited on*, page 80.

[36] Corinna Cortes and Vladimir Vapnik. Support vector machine. *Machine learning*, 20(3):273–297, 1995.

[37] K Dabov, A Foi, V Katkovnik, and K Egiazarian. Image Denoising by Sparse 3-D Transform-Domain Collaborative Filtering. *IEEE Transactions on Image Processing*, 16(8):2080–2095, 2007.

[38] Kostadin Dabov, Alessandro Foi, Vladimir Katkovnik, and Karen Egiazarian. Image denoising with block-matching and 3d filtering. In *Electronic Imaging 2006*, pages 606414–606414. International Society for Optics and Photonics, 2006.

[39] Jeffrey Dean and Sanjay Ghemawat. Mapreduce: simplified data processing on large clusters. *Communications of the ACM*, 51(1):107–113, 2008.

[40] Rachid Deriche. Using canny's criteria to derive a recursively implemented optimal edge detector. *International journal of computer vision*, 1(2):167–187, 1987.

[41] J. Dongarra, H.E. Bal, M. Hamilton, and Y. Legre. Keynotes. In *High Performance Computing Simulation (HPCS), 2015 International Conference on*, pages 1–5, July 2015.

[42] Rakesh Dugad and UDAY B Desai. A tutorial on hidden markov models. *Signal Processing and Artifical Neural Networks Laboratory, Dept of Electrical Engineering, Indian Institute of Technology, Bombay Technical Report No.: SPANN-96.1*, 1996.

[43] R P Eatough, N Molkenthin, M Kramer, A Noutsos, M J Keith, B W Stappers, and A G Lyne. Selection of radio pulsar candidates using artificial neural networks. *arXiv.org*, (4):2443–2450, May 2010.

[44] PP Ewald. Introduction to the dynamical theory of x-ray diffraction. *Acta Crystallographica Section A: Crystal Physics, Diffraction, Theoretical and General Crystallography*, 25(1):103–108, 1969.

[45] Mike Folk, Gerd Heber, Quincey Koziol, Elena Pourmal, and Dana Robinson. An overview of the HDF5 technology suite and its applications. *EDBT/ICDT Array Databases Workshop*, pages 36–47, 2011.

[46] Gianluca Geloni, E Saldin, L Samoylova, E Schneidmiller, H Sinn, Th Tschentscher, and M Yurkov. Coherence properties of the european xfel. *New Journal of Physics*, 12(3):035021, 2010.

[47] P Hart, S Boutet, G CarmI, A Dragone, B Duda, D Freytag, G Haller, R Herbst, S Herrmann, C Kenney, J Morse, M Nordby, J Pines, N van Bakel, M Weaver, and G Williams. The Cornell-SLAC pixel array detector at LCLS. In *Nuclear Science Symposium and Medical Imaging Conference (NSS/MIC), 2012 IEEE*, pages 538–541. IEEE, 2012.

[48] Simon Haykin and Neural Network. A comprehensive foundation. *Neural Networks*, 2(2004), 2004.

[49] Jeff Heaton. Encog: Library of interchangeable machine learning models for java and c#. *Journal of Machine Learning Research*, 16:1243–1247, 2015.

[50] Robert Hecht-Nielsen. Theory of the backpropagation neural network. In *Neural Networks, 1989. IJCNN., International Joint Conference on*, pages 593–605. IEEE, 1989.

[51] Richard Henderson. The potential and limitations of neutrons, electrons and x-rays for atomic resolution microscopy of unstained biological molecules. *Quarterly reviews of biophysics*, 28(02):171–193, 1995.

[52] Christian Igel and Michael Hüsken. Improving the rprop learning algorithm, 2000.

[53] Linda C Johansson, David Arnlund, Gergely Katona, Thomas A White, Anton Barty, Daniel P DePonte, Robert L Shoeman, Cecilia Wickstrand, Amit Sharma, Garth J Williams, et al. Structure of a photosynthetic reaction centre determined by serial femtosecond crystallography. *Nature communications*, 4, 2013.

[54] Stephen Kaisler, Frank Armour, J Alberto Espinosa, and William Money. Big Data: Issues and Challenges Moving Forward. In *2013 46th Hawaii International Conference on System Sciences (HICSS)*, pages 995–1004. IEEE, December 2012.

[55] Vasiliki Kalavri and Vladimir Vlassov. Mapreduce: Limitations, optimizations and open issues. In *Trust, Security and Privacy in Computing and Communications (TrustCom), 2013 12th IEEE International Conference on*, pages 1031–1038. IEEE, 2013.

[56] Robert Klanner, Julian Becker, Eckhart Fretwurst, Ioana Pintilie, Thomas Pöhlsen, Joern Schwandt, and Jiaguo Zhang. Challenges for silicon pixel sensors at the european xfel. *Nuclear Instruments and Methods in Physics Research Section A: Accelerators, Spectrometers, Detectors and Associated Equipment*, 730:2–7, 2013.

[57] G ROBERTS Lawrence. Machine perception of three-dimensional solids. *Ph. D. Thesis*, 1963.

[58] Quoc V Le, Marc'Aurelio Ranzato, Rajat Monga, Matthieu Devin, Kai Chen, Greg S Corrado, Jeff Dean, and Andrew Y Ng. Building high-level features using large scale unsupervised learning. *arXiv.org*, December 2011.

[59] Marc Lebrun. An Analysis and Implementation of the BM3D Image Denoising Method. *IPOL Journal ()*, 2:175–213, 2012.

[60] Marc Lebrun. An analysis and implementation of the bm3d image denoising method. *Image Processing On Line*, (2012), 2012.

[61] Erik Lindholm, John Nickolls, Stuart Oberman, and John Montrym. Nvidia tesla: A unified graphics and computing architecture. *IEEE micro*, (2):39–55, 2008.

[62] Wei Liu, Daniel Wacker, Cornelius Gati, Gye Won Han, Daniel James, Dingjie Wang, Garrett Nelson, Uwe Weierstall, Vsevolod Katritch, Anton Barty, et al. Serial femtosecond crystallography of g protein–coupled receptors. *Science*, 342(6165):1521–1524, 2013.

[63] Andrew McAfee, Erik Brynjolfsson, Thomas H Davenport, DJ Patil, and Dominic Barton. Big data. *The management revolution. Harvard Bus Rev*, 90(10):61–67, 2012.

[64] Uwe Meyer-Baese and U Meyer-Baese. *Digital signal processing with field programmable gate arrays*, volume 65. Springer, 2007.

[65] Claudio Nicolini and Eugenia Pechkova. Nanocrystallography: an emerging technology for structural proteomics. *Expert review of proteomics*, 1(3):253–256, 2004.

[66] Tohru Nitta. Solving the XOR problem and the detection of symmetry using a single complex-valued neuron. *Neural Networks ()*, 16(8):1101–1105, 2003.

[67] Mark Nixon, Mark S Nixon, and Alberto S Aguado. *Feature extraction & image processing for computer vision*. Academic Press, 2012.

[68] Jeffrey M Perkel. Decoding protein structure, one femtosecond at a time, 2014.

[69] Stefania Perri, Marco Lanuzza, Pasquale Corsonello, and Giuseppe Cocorullo. A high-performance fully reconfigurable FPGA-based 2D convolution processor. *Microprocessors and Microsystems ()*, pages 381–391, 2005.

[70] Dean A Pomerleau. Efficient training of artificial neural networks for autonomous navigation. *Neural Computation*, 3(1):88–97, 1991.

[71] Judith MS Prewitt. Object enhancement and extraction. *Picture processing and Psychopictorics*, 10(1):15–19, 1970.

[72] Lars Redecke, Karol Nass, Daniel P DePonte, Thomas A White, Dirk Rehders, Anton Barty, Francesco Stellato, Mengning Liang, Thomas RM Barends, Sébastien Boutet, et al. Natively inhibited trypanosoma brucei cathepsin b structure determined by using an x-ray laser. *Science*, 339(6116):227–230, 2013.

[73] M Reed and B Simon. Methods of modern mathematical physics. vol. ii, fourier analysis. *Self-Adjointness Academic Press, San Diego*, 1975.

[74] James Reinders. An overview of programming for intel xeon processors and intel xeon phi coprocessors. *Intel Corporation, Santa Clara*, 2012.

[75] Martin Riedmiller. Advanced supervised learning in multi-layer perceptrons - from backpropagation to adaptive learning algorithms, 1994.

[76] Martin Riedmiller and I. Rprop. Rprop - description and implementation details, 1994.

[77] I K Robinson and D J Tweet. Surface x-ray diffraction. *Reports on Progress in Physics*, 55(5):599, 1992.

[78] Raúl Rojas. *Neural networks: a systematic introduction*. Springer, 1996.

[79] Lucas M Russo, Emerson C Pedrino, Edilson Kato, and Valentin Obac Roda. *Image convolution processing: A GPU versus FPGA comparison*. IEEE, 2012.

[80] Matthew Scarpino. Opencl in action: How to accelerate graphics and computation. ny, 2012.

[81] Hanno Scharr. *Optimal operators in digital image processing*. PhD thesis, 2000.

[82] Benjamin Felix Schock. Cross application communication on numa-systems. Master's thesis, HTW Berlin, 2014.

[83] GK Shenoy and J Stoehr. Lcls-the first experiments. *SLAC, Stanford*, 2000.

[84] MS Smyth and JHJ Martin. x ray crystallography. *Journal of Clinical Pathology*, 53(1):8, 2000.

[85] Irwin Sobel and Gary Feldman. A 3x3 isotropic gradient operator for image processing. *a talk at the Stanford Artificial Project in*, pages 271–272, 1968.

[86] J Sola and J Sevilla. Importance of input data normalization for the application of neural networks to complex industrial problems. *IEEE Transactions on Nuclear Science*, 44(3):1464–1468, 1997.

[87] John E Stone, David Gohara, and Guochun Shi. Opencl: A parallel programming standard for heterogeneous computing systems. *Computing in science & engineering*, 12(1-3):66–73, 2010.

[88] Isa Servan Uzun, Abbes Amira, and Ahmed Bouridane. Fpga implementations of fast fourier transforms for real-time signal and image processing. In *Vision, Image and Signal Processing, IEE Proceedings-*, volume 152, pages 283–296. IET, 2005.

[89] Jos van Wezel, Achim Streit, Christopher Jung, Rainer Stotzka, Silke Halstenberg, Fabian Rigoll, Ariel Garcia, Andreas Heiss, Kilian Schwarz, Martin Gasthuber, et al. Data life cycle labs, a new concept to support data-intensive science. *arXiv preprint arXiv:1212.5596*, 2012.

[90] Dingjie Wang. *Methods and Instrumentation of Sample Injection for XFEL Experiments*. PhD thesis, Arizona State University, 2014.

[91] Paul J Werbos. Backpropagation through time: what it does and how to do it. *Proceedings of the IEEE*, 78(10):1550–1560, 1990.

[92] Thomas A White, Richard A Kirian, Andrew V Martin, Andrew Aquila, Karol Nass, Anton Barty, and Henry N Chapman. CrystFEL: a software suite for snapshot serial crystallography. *Journal of Applied Crystallography*, 45(2):335–341, March 2012.

[93] Tom White. *Hadoop: The definitive guide*. " O'Reilly Media, Inc.", 2012.

[94] Arthur James Cochran Wilson. *Elements of X-ray Crystallography*. Addison-Wesley Reading, Massachusetts, 1970.

[95] Matei Zaharia, Mosharaf Chowdhury, Michael J Franklin, Scott Shenker, and Ion Stoica. Spark: cluster computing with working sets. In *Proceedings of the 2nd USENIX conference on Hot topics in cloud computing*, volume 10, page 10, 2010.

[96] Zepu Zhang, Nicholas K. Sauter, Henry van den Bedem, Gyorgy Snell, and Ashley M. Deacon. Automated diffraction image analysis and spot searching for high-throughput crystal screening. *Journal of Applied Crystallography*, 39(1):112–119, January 2006.